Robert Ells

A history of New Brunswick geology

A HISTORY

OF

NEW BRUNSWICK GEOLOGY.

BY

R. W. ELLS, M.A.,
Of the Geological Survey of Canada.

MONTREAL:
GAZETTE PRINTING COMPANY.

1887.

Robert Ells

A history of New Brunswick geology

ISBN/EAN: 9783337204440

Printed in Europe, USA, Canada, Australia, Japan

Cover: Foto ©berggeist007 / pixelio.de

More available books at **www.hansebooks.com**

A HISTORY

OF

NEW BRUNSWICK GEOLOGY.

BY

R. W. ELLS, M.A.,
Of the Geological Survey of Canada.

MONTREAL:
GAZETTE PRINTING COMPANY.
1887.

A HISTORY

OF

NEW BRUNSWICK GEOLOGY,

BY

R. W. ELLS, M.A.,

Of the Geological Survey of Canada.

The literature pertaining to the Geology of New Brunswick, embracing a period of nearly half a century, and expressing the views of nearly a score of observers, has become at length so voluminous as to be in a manner somewhat unintelligible to one not familiar with the Province and the peculiar geological problems there presented. For this reason it has been deemed advisable to produce, in concise form, an epitome of the work done and the opinions held, from time to time, by those who have labored in this field; especially when we consider that, no matter how interesting the subject, but few persons have the time necessary for the careful perusal of all the reports bearing on this special field, many of which also are now not to be obtained, having been long out of print.

The writer can only plead, in excuse for this attempt, a somewhat intimate acquaintance with the topographical and geological features of the Province, extending over a period of some ten years devoted to the study of its geology.

The systematic study of the geological structure of New Brunswick may be said to have begun with the appointment, by the Local Government, to the position of Provincial Geologist in 1838, of Dr. A. Gesner, a man of undoubted ability. The results of his labors were presented in a series of reports, which may be said to constitute the basis of our knowledge in this direction. The first of these, appearing in that year, embraced the results of his examination of the country bordering on the Bay of Fundy, west of the St. John River, and along that stream as far up as Fredericton. This was followed, in 1839, by a report on the coast district east of St. John, extending to the head of Shepody Bay, with a brief description of

Dr. A. Gesner, 1838-43.

the Tormentine peninsula and the country along the Hammond River. His third season's work embraced generally the counties of St. John and Kings, with a very complete description of the Grand Lake coal-field. The fourth described the western portion of the province as far north as Woodstock, on the River St. John, with the country contiguous to that river, as well as the character of the coal-fields lying along the Main South-West Miramichi River, while the fifth included generally the area lying north of a line extending from Woodstock to Bathurst, on the Bay of Chaleurs.

Considering the state of geological science in that early day, the reports of Dr. Gesner contained a large amount of valuable information. It must be borne in mind that the nomenclature of the science was very limited. The grand formation which now comprises the fundamental rocks of our record, the Laurentian, had not then received its now world-wide designation, nor had the term Huronian been even thought of. The divisions into Primary and Secondary, Old and New Red Sandstone, Coal measures and granitic rocks, composed the bulk of the geological scale.

There could, therefore, have been no attempt to separate the rocks of the older systems into such an arrangement as now exists. But Gesner evidently did a large amount of good work in the delineation of his areas of Coal measures and New Red Sandstones, his Transition or slate and limestone group, and his volcanic rocks, though many of his boundaries were, of necessity, from the sparsely settled and, in consequence, comparatively inaccessible character of the country, far from correctly laid down, and the stratigraphical order, as given in his reports, is, in some cases, the reverse of what is now known to be the true position.

He pointed out also the presence of the two great areas of granitic rocks which traverse the province, one along its southern portion, the other diagonally across the northern half, extending south-westerly into the State of Maine, where these two areas evidently unite. They were considered by him as of Primary age, and included a large proportion of the felsitic rocks, with which, in some places, the granites are intimately connected. He held that these so-called Primary ridges were flanked by beds of Cambrian age, consisting of slates and hard-grained sandstone, styled by him greywacke, while to the north of the northerly belt the great Silurian fossiliferous area of slates and limestones was clearly indicated. Then, as now, the geology of the southern part of the province was found to be much more complicated than that of the northern portion. The igneous rocks were arranged into two belts, the one composed of true granite and syenite, with mica or hornblende, the representatives of the red granitic areas now recognized in Charlotte county and western Kings ; the other comprised a large portion of the old pre-Cambrian syenites and felsites of the present day, and were regarded as intrusive and as overflowing the schistose strata with which they are associated.

Resting upon the flanks of these intrusive ridges were two great series, one containing the limestones now regarded as of Laurentian age, together with certain slates about St. John, and classed as the lower series ; another portion, consisting of sandstones and slates and holding fossil plants and tree stems, was recognized as an upper or newer division. These two groups, the lower of which was supposed to belong to the Silurian system, were held to pertain to the greywacke or transition series and to overlie another area of schists, sandstones, conglomerates, etc., unconformably, which, from an apparent absence of organic remains, was styled Primary.

Concerning the rocks which underlie directly the great central Carboniferous area, the limestones were correctly placed in the lower portion of that system, but the associated red sandstone, conglomerates and marls, together with similar rocks in the Valley of the Kennebacasis and Petitcodiac Rivers, were regarded as of more recent age and referred to the horizon of the New Red Sandstone ; while other red sediments on the St. John River, near Hampstead, were regarded as older and of the age of the Old Red Sandstone. The areas of soft red sandstones and shales along the Bay of Fundy, at Quaco and at other points east of St. John, were also regarded as of New Red Sandstone age.

Reviewing the reports of Dr. Gesner, one can hardly fail to be impressed with his evident desire to convince the Government, and through it the people, that the mineral resources of the Province were practically limitless. This is more particularly the case in regard to the central coal basin and the iron ore deposits of Kings and Queens counties ; and this feature has been ably criticized by the late Dr. Robb in his subsequent report on the coal fields of the Province, published in the report of Prof. Johnson on the agricultural resources of New Brunswick, in which he clearly points out the unwarrantable exaggerations of Dr. Gesner as to the boundless stores of mineral wealth.

Doubtless many false hopes were raised by this unwise policy, and
ground was given for much unprofitable controversy, before the true
geological relations of the Carboniferous rocks were finally established. To his researches, however, we must ascribe the discovery
of that wonderfully rich mineral deposit called generally "Albert
coal," and concerning the true character and composition of which
he appears to have had a just judgment, regarding it as an altered
asphalt in the face of the combined opposition of the majority of the
leading scientists of the day.

It is unfortunate that, although Dr. Gesner spent the greater
part of five years in his preliminary surveys of this Province, his
labors ended without the publication of any geological map which
might embody the result of the very large amount of exploratory
work he evidently accomplished ; and this is the more to be regret-
ted, since many of the points so graphically described by him, lose, in
consequence, very much of their actual value to the general reader.

Dr. J. Robb,
1849-50.

Following Dr. Gesner, the next writer on the subject was the late
Dr. Robb, of Kings College, Fredericton (now the University of
New Brunswick), who, in 1849-50, published a geological map
which is embodied in the report of Prof. Johnson, already referred to,
and contributed a chapter to that work. Comparing this map with
the reports just described, it will be seen that a considerable advance
has been made both as regards stratigraphy and nomenclature. The
Primary and Transition groups are now arranged under the head of
Cambrian and Lower and Upper Silurian, but no distinction was
made in the red-colored sediments, the whole being massed under
one heading, though comprising areas which range from the Devonian to the top of the Upper Carboniferous. He, however, clearly
understood the true position of the Lower Carboniferous sandstones
and conglomerates as underlying the Coal measures—to which
arrangement Dr. Gesner dissented.

The belts of granite, both of the northern and southern areas, were
indicated roughly, but much of what is now called pre-Cambrian,
embracing a great thickness of volcanic rocks, was included under
the head of traps, syenites, felspar-porphyries, etc. ; while the Cambrian rocks were supposed to include, not so much the recognized
Cambrian of the present day, as what are undoubtedly the oldest of
the pre-Cambrian rocks, viz., the Laurentian limestones, syenites
and gneisses, with associated slates. The limits of the central
Carboniferous basin were outlined with considerable correctness,
and the various coal crops clearly defined.

Dr. Robb also seems to have perfectly understood the general unproductivevess of this area, and to have vigorously confuted the exaggerated statements previously made concerning its economic importance.

In the northern portion of the Province, the boundaries of the great Silurian area north of the Tobique River, in so far as accessible, were quite correctly delineated, though, as communication was necessarily difficult, the various geological features were otherwise, for a great part of the area, largely conjectural.

Following the publications of Dr. Robb, the next papers on the subject are brief articles by Messrs. Jackson and Taylor, bearing principally upon the disputed mineral of the Albert Mine. These appeared in 1851, and were succeeded in 1852 by another, published in the *Geological Journal*, London, on the structure and geological relations of this famous deposit, by Mr. J. W., now Sir William Dawson.

Perley's Handbook for Emigrants, while containing some information concerning the geological structure of the Province, can scarcely be said to have advanced our knowledge greatly ; since the remarks there contained, were taken largely from the works of previous writers. But in 1855, the first edition of the Acadian Geology contained two chapters relating to this Province, the former of which, pertaining to the Carboniferous system, is a valuable contribution, and the conclusions there expressed as to the horizon of the central area, as well as the divisions of the south-eastern portion in Albert county, have been very fully confirmed by the most recent investigations.

Concerning the older series of rocks developed about St. John, sufficient work had not at that time been done to determine their true age. The richly fossiliferous Primordial or Cambrian slates of that city and of the Hammond River valley to the eastward had not then been studied; and only a few imperfect remains had been found, whose age could not be ascertained. As a consequence, the pre-Cambrian age of the rocks which unconformably underlie these at many points could not be established, while opportunities for comparison between the crystalline portion and the recognized Laurentian and Huronian of Ontario and Quebec, which, through the labors of Logan, Murray and Hunt, had now come into great prominence, were not sufficient to establish their true position on lithological grounds. The conclusions on this area then stated, in default of extended personal observations, were therefore of necessity largely

(margin notes)
Jackson and Taylor, 1851.

Sir William Dawson's Acadian Geology, 1st edition, 1855.

those of Dr. Robb, who, at that time, and, in fact, for some years later, was regarded as the standard authority on New Brunswick geology.

During the succeeding years to 1860, but little geological work appears to have been done, except in a desultory way; but about that date, several young gentlemen of St. John, notably Messrs.

Messrs. Hartt, Bailey, and G. F. and C. R. Matthew, 1860.

Hartt and the brothers G. F. and C. R. Matthew, began the careful systematic study of the rock formations about that city. They were ably assisted by Prof. L. W. Bailey, who had lately been appointed to the chair of Natural History in the University of New Brunswick, which had become vacant through the death of Dr. Robb. These gentlemen were fortunate in discovering a rich fauna and flora in many of the beds, both in the city itself as well as at various points to the east and west. The collected fossils were submitted to Sir J. W. Dawson, the principal of McGill College, and the recognized authority on fossil botany, who speedily determined the horizon of a portion of the strata as Devonian, while the fauna of another portion clearly appertained to a much older period. The results of these examinations were first made public in a paper read before the

Sir William Dawson, 1861.

Natural History Society of Montreal, in 1861, and the Devonian character of the flora fully stated. In the same paper, however, the author stated that the associated slates and limestones, regarded by Robb as Cambrian, but whose relations had not been clearly made out, might possibly belong, on stratigraphical grounds, to the Devonian or Silurian. These results were also communicated, in a much more extended form, in a paper to the Geological Society of London,

Sir William Dawson, 1862.

published in 1862 ; in which also it was stated that the Devonian might possibly include what is now known to be Primordial, as well as the Laurentian gneisses, limestones and associated rocks.

It will thus be seen that the geology of this section was an exceedingly difficult problem to decipher and one requiring great stratigraphical skill, as well as extensive knowledge of the obscure fossil forms that were being discovered from time to time, the horizon of which had not yet been accurately determined ; and it was not until several years later that light began to break in upon this intricate question ; for in the next publication on this subject, which was

Mr. G. F. Matthew, 1863.

a communication from Mr. G. F. Matthew, to the Canadian Naturalist in 1863, while re-casting the groups, as stated by Principal Dawson, in the previous year, and giving them local names, he failed to see any real ground for the separation of the metamorphic

portion, comprising the limestones, gneisses, &c., from the fossil
iferous Devonian. He, however, gave the name of " Portland," to Portland group.
the lowest members of the group, from the fact of their being extensively developed in that suburb of St. John, and assigned them to
the horizon of the lower Devonian, or possibly the upper part of the
Silurian system.

To an apparently overlying series, embracing a considerable
thickness of greenish grey slates, red, slaty conglomerates and shales,
with red conglomerates, grits and hard grey sandstone, the name
" Coldbrook " was given, while a third division, styled the " St. John Coldbrook
group," comprised a series of dark grey slates and sandstones, to a group.
great extent the Primordial of the present day, in which were found
a lingula and several other obscure fossils.

His succeeding group, the " Bloomsbury," while largely of volcanic Bloomsbury
origin, included in its upper part some five hundred feet of slates group.
and conglomerates, apparently devoid of fossils ; overlying which
came the Little River and Mispec groups, largely composed of sand- Little River
stones, slates and conglomerates of various colors, but distinguished group.
and Mispec
group.
throughout by a great abundance of fossils, principally plants,
though comprising also crustaceans and the remains of insects.

The apparent difficulty of separating the metamorphic portion from
the fossiliferous and recognized Devonian, arose in great measure
from the seeming interstratification of the various groups, and
the complicated structure arising from the infolding of newer strata
with those of great antiquity, due to a series of anticlinals which are
often completely overturned, in conjunction with profound faults ;
and it was not till the researches of another year by those indefatig-
able workers, Messrs. G. F. and C. R. Matthew, that the beds which First discovery
the former gentleman had designated the St. John group, and which of Primordial
fossils, 1862-63,
had for a long time been known to hold certain obscure traces of by Messrs.
Hartt and
organic life, disclosed a new and entirely unexpected suite of fossils ; Matthew.
these included brachiopods and trilobites, the latter in considerable
variety, but generally of small sizes, which, on comparison with the
published memoirs of Barrande on the Primordial of Europe, resulted
in the establishment of a similar Primordial zone in New Brunswick,
a discovery, the importance of which in relation to the elucidation of
the geology of this section, can scarcely be overrated.

The credit of solving this difficult problem is largely due to the C. F. Hartt on
late Mr. C. F. Hartt, one of the earliest and most zealous workers in Primordial
fossils, 1864.
this field, who at that time was engaged in the Museum of Compara-

tive Zoology in Harvard College. The results of his investigations on the fossils collected by the Messrs. Matthew in 1862, and by himself in the following year, were communicated in a preliminary paper to Prof. Bailey in 1864, and published in the " Observations on the Geology of Southern New Brunswick " by that author in 1865.

The discovery of the Primordial or Cambrian zone in the vicinity of St. John worked an entire revolution in the stratigraphy of that portion of the province. The puzzling admixture of the most highly metamorphic rocks with those entirely unaltered and abounding in fossils was again investigated, and by the aid of the new light, a tolerably clear knowledge of the structure was obtained. The map accompanying the report just referred to, as compared with the hitherto recognized standard of Dr. Robb, shows many important changes in the various geological formations   One of the first corrections made was the transposition of the crystalline limestones gneisses, &c., from their former doubtful position to their proper

Pre-Cambrian first established place below the Cambrian, which was found to overlie them unconformably at various points ; while their resemblance to recognized Laurentian rocks of Ontario and Quebec, which had already been pointed out by Sir William Dawson, rendered it exceedingly probable that they might occupy a similar position in the geological record. They were, therefore, so arranged and have since been regarded by most geologists as among the fundamental rocks of the province.

The lithological characters of the various rocks which compose these Azoic or Eozoic strata have been given in different papers on the subject. Their high degree of metamorphism was commented

Sir William Dawson, 1861. on by Sir William Dawson as early as 1861, and the descriptions of the several divisions of slates, limestones, quartzites, syenites, and gneiss were subsequently stated in Prof. Bailey's report in 1865, under the heading of the " Portland group, " and will be considered later.

Prof. L. W. Bailey, 1864, northern New Brunswick. Following the paper of Mr. Matthew in 1863, on the rocks in the vicinity of St. John, appeared one by Prof. L. W. Bailey in the Canadian Naturalist, detailing observations made in 1863 during a canoe voyage across the northern part of the Province, by way of the Tobique and Nipisiguit rivers. This paper is of special interest as giving us the first scientific account of the geology and botany of the country along these streams, since on the map of Dr. Robb, while

the areas of Upper Silurian and Lower Carboniferous on the Tobique
were laid down with tolerable accuracy, the boundaries of the formations on the other streams flowing east to the Bay of Chaleurs were
largely imaginary. The great areas of felsites and other crystalline
rocks about the head waters of the Nipisiguit were noted by Prof.
Bailey, though their age was not determined, and the Lower Silurian
aspect of the strata on the lower portion of this stream was pointed
out. This was followed in the same year, 1864, by his " Notes on L. W. Bailey,
the mineral resources of New Brunswick," which, while giving noth- resources of
ing specially new in reference to the geology, contained much inter- New Brunswick
esting matter concerning the mining industries, then in their infancy.
It was in turn followed the next year, 1865, by two reports on the L. W. Bailey,
Geology of the Province, the first by Prof. Bailey on the southern southern New
portion already alluded to, and the second by Prof. H. Y. Hind, also Brunswick,
on its general geology, but more especially of interest in reference to Geol. of New
the northern portion, and containing much information on the mine- Brunswick,
rals of economic value. In both of these a marked increase in the
nomenclature of the science is manifest. The researches of Prof.
Hartt on the Primordial already noted, led, not only to the separation of a portion of the crystalline rocks as Laurentian, but made a still
further stride by the removal of a second portion, lying stratigraphically between the Laurentian and the Cambrian, which was erected
into a distinct group with the title of Coldbrook, and assigned to the
Huronian system. This division was in time subdivided into an
upper and lower, the former of which, consisting largely of reddish
strata, was regarded as of purely sedimentary character, while the
latter, composed largely of hard, greenish rocks, was held to be
chiefly of volcanic origin, the thickness of the whole being estimated
at 5,000 feet.

The recognized areas of Huronian rocks were, however, as yet very
limited. The intricate stratigraphy of the south coast still prevented
the separation of much of what is now known to be of that age from
the position it had so long held as presumably Devonian or Silurian.
From their apparently interstratified position among the plant
bearing beds of the former, at several points east of St. John, it was
inferred that a great thickness of strata, highly metamorphic in
character, and now known to be among the oldest in the Province,
constituted an integral portion of that series.

This belt which has an extensive development in eastern St. John,
Kings and Albert counties and which has since been found to un-

conformably underlie rocks of Primordial age, had been brought into
its apparent abnormal superposition upon the fossiliferous Devonian
by a complicated system of faults and overturns, which at that time
had not been investigated.

Kingston
group

Another great group of rocks, also for the most part highly metamorphic, and designated by the term " Kingston," was brought prominently into notice in the same report. More difficult, apparently,
of location than even the Coldbrook, its exact position could not,
at that time, be determined by the New Brunswick geologists, and
a suite of specimens was accordingly submitted to the inspection of
Sir W. Dawson and Dr. Hunt of Montreal.

The lithological characters of the group were stated to be very
like those found in the rocks of the Cobequid series of Nova Scotia,
which were then regarded as of Upper and Middle Silurian age, but
it was also stated that possibly portions of the group might pertain
to the Devonian system, from their resemblance to the supposed
Devonian of south eastern New Brunswick, while the similarity of
many of the rocks to older or true Plutonic masses was also pointed
out. The true position of this group of rocks, which has a considerable development in the southern part of the province, will be
considered later.

The Devonian of this area, as described in the report in question,
has, since its publication, been greatly modified. It became very
evident, as the relations of the Huronian and Primordial were more

Division of the
Bloomsbury
group.

fully understood, that a considerable thickness of what was styled the
Bloomsbury group, which represented the lower portion of the Devonian, possessed lithological characters very similar to much of what
was now called Coldbrook, and that it was clearly separable into two
parts, the upper of which only,—embracing some 500 feet of sandstones, shales and conglomerates,—was referable to the Devonian;
the lower portion, which was largely volcanic, being transferred to

Division of
Little River
group into
Dadoxylon
sandstone and
Cordaite shale.

the pre-Cambrian.

The Little River group, the second division, was also clearly
separable into two, the Dadoxylon sandstone and the Cordaite
shales; the former of which, consisting for the most part of hard
grey sandstones, shales and grits, and representing a total thickness
of about 2,800 feet, was characterized throughout by a wonderfully
rich and important flora, of which large collections were made and
carefully determined by Sir W. Dawson.

The second subdivision of the Little River group was, in 1865,
held to embrace rocks of very dissimilar character. In addition to

fine shales and sandstones, holding the remains of Cordaites from which the formation received its name, it was thought that a great thickness of metamorphic rocks, chloritic and talcose schists, felsites, etc., presenting a marked resemblance to strata, which at a later period were recognized as Huronian, constituted, from their position upon the Dadoxylon sandstone, an upper part of the same series.

This view was entertained for some years, or till the subsequent study of the many faults seen on the east side of the St. John harbor, and which affected this area very seriously, disclosed the fact that the apparent position of the metamorphic upon the unaltered portion was due to an overturn, or possibly to a sliding fault. The error thus stated in 1865 and repeated by Prof. Hind in his report in the same year, was again reproduced in the geological map which accompanied the second edition of the Acadian Geology in 1868; but it was soon afterwards discovered by the local geologists, and the metamorphic portion was separated, and established as a division of the Huronian, under the title of the " Coastal group." *Separation of the Coastal group.*

The highest beds of Devonian age were arranged under the head of the Mispec group. They consisted principally of reddish con- *Mispec group.* glomerates and slates, the former holding fragments of felsitic rocks, red sandstones and slaty limestones in a reddish slaty paste. These were not found to contain fossils.

Passing to the consideration of the next system, we find the gene- *Carboniferous system.* ral outlines of the Carboniferous indicated with tolerable accuracy. The observations appear, however, to be largely directed to the structure of the lower division ; which, more especially in its eastern extension, included considerable areas of the middle Carboniferous ; sufficient opportunities for study not having been afforded in this direction to admit of their complete separation. The various subdivisions of the Lower Carboniferous, as there laid down for the eastern part of Albert county, were verified on subsequent detailed examination, though certain sections elsewhere appeared at first view to present a somewhat different arrangement ; more especially in regard to the true position of the bituminous or " Albert shales."

In its distribution, the Lower Carboniferous formation was found to constitute a well defined belt, underlying, throughout its whole extent, the Middle Carboniferous basin of the interior. It also comprised certain areas in eastern Albert and Kings counties which rested upon the flanks of what was then regarded as the metamor‹

phic Devonian of the Caledonia Mountain range. West of St. John, however, its presence was not definitely noted, certain rocks in the vicinity of Lepreau village and harbour, which had been regarded by Dr. Gesner as of New Red Sandstone and Carboniferous age, being found to pertain to a much lower horizon, representing the upper part of the Devonian system.

Concerning the structure of the central or Middle Carboniferous basin, but few additional details were added to the views already expressed by the late Dr. Robb. Fossil plants were obtained at points about Grand Lake, as well as in the vicinity of the Miramichi River and the Bay of Chaleurs, which appeared to present what was at that time regarded as a mingling of forms, including some portions even of the Upper Carboniferous, while fossils from other outliers of the formation, distributed along the North side of the Bay of Fundy, indicated a Millstone-grit age. The discovery however by Mr. C. R. Matthew, of the presence of a considerable area of micaceous slates, which are probably Devonian, in the very heart of the Grand Lake coal field, was very important, since it confirmed the view as to the apparent thinness of the measures at this point ; while the fossils in the overlying rocks were of the horizon of the Millstone-grit.

Prof. H. Y.
Hind, 1865.

The report of Prof. H. Y. Hind to the New Brunswick government, (1865) contains a large amount of very valuable and interesting matter, relating, not only to the geology, but to the mineral and agricultural resources of the province. A marked advance in geological knowledge is evident in connection with the work, more especially in regard to the northern portion, cencerning which area, our information up to that time, owing to its largely inaccessible character, was very limited. A number of sections were made by Prof. Hind along the various streams of the interior. The southern outline of the Upper Silurian was more clearly defined, both along the coast of the Bay of Chaleurs and on the Upsalquitch and Tobique rivers, as well as on the St. John, to the west. The great belt of metamorphic rocks of the interior, crossed by Prof. Bailey on the Nipisiguit, were examined and classed by Prof. Hind, principally on lithological grounds, as the equivalents of the Quebec Group of Canada, as laid down by the late Sir William Logan. The great areas of granite and syenite, both of the southern and northern portions of the province, were held to be intrusive and of Devonian age. No specially new facts bearing on the geology of the southern part of the province

were advanced; the views of Bailey and Matthew, in regard to the
Devonian age of the metamorphic rocks of Albert and Kings, being
assented to. In the western area, however, the partly metamorphic
belt underlying the Carboniferous basin on its northwest side, and
containing the Antimony ores of Prince William,—rocks which had
been regarded by Drs. Robb and Gesner as Cambrian, and by
Matthew (see Appendix D, Report 1865,) as Lower Silurian—was
also classed by Hind in his Quebec Group, and paralleled with the
upper and slaty members seen on the Nipisiguit and other rivers in
the northern part of the province.

On the lower Restigouche, between Campbellton and Dalhousie,
Devonian rocks were recognized, skirting the shores, and associated
with traps which present conspicuous hill features in this locality.
Small outliers of Carboniferous conglomerate, near Dalhousie station,
were also indicated, and their true horizon was stated, in contradis-
tinction to the early views of Dr. Gesner, who had regarded them as
of the age of the Carboniferous and New Red Sandstone.

Following the report of Prof. Hind, the next contribution of Sir William
note to the geology of the province was contained in the second Acadian
editon of the Acadian Geology, 1868. The observations there given 1868.
were based largely on the work of Dr. Robb, Prof. Bailey, and Mr.
Matthew. A large amount of additional information, relating principally to palæontological details, was also given, embracing the various
formations from the St. John or "Acadian Group," to the Carbon- Name
iferous, both inclusive. The general distribution of the Laurentian first proposed
and Huronian, the latter of which was still confined to the Coldbrook by Sir William
group, remained the same as in Prof. Bailey's report, 1865. In the
metamorphic rocks which surrounded the central Carboniferous basin,
regarded as Lower Silurian, was included a large area in northern
Charlotte county, and in the southern part of Sunbury and Queens,
which was afterwards found to be of much more recent age. Following
the determinations of Mr. Matthew, the belt extending diagonally
through the province, from its south-west angle to the vicinity of
Bathurst, was also regarded as Lower Silurian, and not far removed
in age from the St. John group. New and valuable matter relating to
the Upper Silurian of Dalhousie and vicinity, with lists of characteristic fossils, was given, while to the same horizon was referred the
Kingston group of the southern part of the province, which was also
again paralleled with the Cobequid series of Nova Scotia. The
accompanying map contained much new information, indicating a

marked increase in our general knowledge of the structure of this interesting field.

In consequence of the passage of the Confederation Act in 1867, New Brunswick, at that date, came, to a certain extent, under the control of the Federal Government ; and, in the ensuing year, the operations of the Geological Survey of Canada, which had been carried on continuously in Ontario and Quebec since 1843, were now extended to the Maritime Provinces. In pursuance of this arrangement, Mr. C. Robb was sent to New Brunswick, to study more closely the structure of the country to the west and north of Fredericton, and along the upper St. John River, while Professor Bailey and Mr. Matthew were assigned work in the southern part of the province. The results of these explorations appeared in two reports by Mr. Robb, in 1869 and 1870, and in a voluminous and exhaustive report by the latter gentlemen in 1871., reviewing the geological history of southern New Brunswick up to that date. This report, which evinces a great amount of painstaking research, has ever since been regarded as a standard work on the geology of this section, although modified in some respects as regards details of structure by later and more systematic explorations.

Beginning of work by Mr. C. Robb in northern and western New Brusnwick, 1868.

Mr. C. Robb, Rep. Geol. Sur. 1860.

The first report by Mr. Robb made no attempt, in so far as related to the formations underlying the Carboniferous system, to decide the actual horizons of the rocks deseribed, with the exception of the more northerly area, already regarded as Upper Silurian. The great metamorphic belt, regarded first as Cambrian and later as Lower Silurian, was divided into two parts, separated by the central granitic area. Their mineralogical and lithological characters were carefully given, and the discovery at one point on a branch of the Nashwaak River, of a band of ochreous slates, holding fossils of Devonian age, apparently intercalated with the metamorphic rocks, led to the supposition that this great formation might be newer than had hitherto been supposed. That the two bands to the north as well as to the south of the granite belt were of the same age, was considered probable. The characters of the great central axis of granite were well described, and the highly altered condition of the various strata in contact with it was pointed out, the rocks in places having assumed a gneissoid structure and containing numerous crystals of staurolite, mica, etc. The extension of the metamorphic belt to the north-east, beyond the Little South-West Miramichi was established, though fossils which might determine their actual age could not at that time be found.

Though no new facts tending to determine more precisely the position of these various groups in the geological scale were adduced in these reports, the large amount of topographical work done enabled Mr. Robb to construct a map of the three counties of York, Carleton and Victoria, upon which his observations of the two seasons were laid down. This resulted in a very great advance over the knowledge already possessed of that region, though the entirely unsettled character of the country east of the upper St. John prevented the accurate delineation of the several outliers of newer rocks, found to exist at various points unconformably overlying the metamorphic series.

In the report by Prof. Bailey and Mr. Matthew, which embraced a detailed account of the structure of the southern part of the province, the Laurentian was divided into two portions, the lower of which consisted largely of gneiss and syenite of greenish, grey and reddish colors with associated masses of diorite. The total thickness of this division was not definitely known. The upper was largely a calcareous series, resting upon the lower at many points, and was made up of crystalline limestone, quartzite and slate, and with limited beds of gneiss. The actual thickness of this division was also somewhat conjectural, but a section on the St. John River, near Indiantown, one of the suburbs of St. John city, gave 1,385 feet, while the lower portion, omitting a considerable thickness of granitoid gneiss, recognized in the western part of the county of St. John, but not seen in the river section, was estimated at 2,850 feet. This would give, for the whole of the Laurentian rocks in this area, a probable thickness of over 5,000 feet. In a series of rocks, however, so extensively faulted and displaced as many of those in the vicinity of St. John, the actual measurement of their thickness is almost an impossibility.

Prof. Bailey and Mr. Matthew on southern New Brunswick, Rep. Geol. Sur. 1870-71.

Assisted by the extensive experience of Dr. T. S. Hunt, at that time an officer of the Geological Survey of Canada, the New Brunswick geologists then proceeded to attack the intricate problem of the metamorphic rocks, which, owing to their complex stratigraphy, had never been satisfactorily arranged.

Upon a more detailed examination, it was conclusively established that rocks of Primordial age were, at several points east of St. John, unconformably placed upon what had so far been regarded as an integral position of the Devonian system while the lithological characters of the latter presented many points of resemblance to those now recognized as of Huronian age. These discoveries,

Huronian rocks—the several groups.

together with the fact that these supposed Devonian rocks were the undoubted continuation of the Coldbrook or Huronian belt of St. John and vicinity, led to their entire removal from the position they had so long held, on grounds both of stratigraphy and lithology, and to their being placed in the Huronian system, under the title of the Coastal group.

The importance of lithological characters being now clearly recognized in the determination of certain horizons, especially when applied to areas of limited extent, attention was next directed to the other great belt of metamorphic rocks which, under the title of the Kingston group, had, up to this time, been regarded as presumably of Upper Silurian age. This position had been assigned it on very much the same grounds that had affected the earlier position of the Coastal, viz. : the supposed interstratification of its equivalents with fossiliferous Upper Silurian sediments which had been observed in eastern Maine. Subsequent examinations, however, proved the stragraphical relations of the true Kingston group to be entirely different. It was found at several points to underlie areas of Primordial slates and, therefore, of necessity, to belong to a much lower position than had been supposed. At the same time, the marked resemblance of the various rocks which composed the group to those of the Coastal and Coldbrook divisions was such that it was deemed most in accordance with the evidence at hand to include it also in the Huronian system. Thus, three great areas of metamorphic rocks, all presumably newer than Laurentian, had been established, though their relative positions to each other had not as yet been clearly defined. The Coldbrook group had, however, been divided into two portions, a lower, including diorites with felsitic and chloritic rock, largely a volcanic series, and an upper, consisting of argillites, sandstones and conglomerates, generally of reddish colors, which was considered to constitute the lowest members of the St. John group.

Since the arrangement of the formations, as given in the report now under discussion (1870-71), has been made, to a large extent, the groundwork of the subsequent publications on the geology of southern New Brunswick, the consideration of the groups newer than the Huronian may here be profitably taken up. Of these, the first Primordial, or in ascending order is the Primordial Silurian, Acadian, or St. John St. John group. group, which latter name it received when regarded as an integral portion of the Devonian system, from the fact of its being largely developed and first studied in and near the city of St. John. By the fortunate recognition of its contained fossils in 1863-64, Prof. Hartt, as already stated, first established the position of the hitherto doubtful metamorphic rocks as pre-Cambrian, and thus afforded a satisfactory basis for future geological work. Six areas were now recognized as probably belonging to the Primordial horizon ; concerning the majority of which no doubt could be entertained, viz.:—

1. St. John and vicinity, with its extension eastward for nearly thirty miles along the valley of the Loch Lomond lakes, continuously, till it reaches the Hammond River.

2. Limited outcrops about the lower part of Kennebecasis Bay, seen on Long and Kennebecasis islands, and on Milkish Head, as well as on the south shore of the bay itself.

3. The valley of the Long Reach of the St. John River. In these three areas, typical fossils are abundant.

4. An area of dark grey and black shales in Wickham, Queens county, resembling in lithological characters the rocks of the previous divisions, but in which no fossils could be detected.

5. A similar series of beds in the valley of the Nerepis River, likewise apparently devoid of organic remains.

6. An area in Charlotte county, near the head of Oak Bay. Other areas of undoubted Primordial age have, however, since been found.

In all these the lithological characters are exceedingly similar, but later explorations on the areas in question led, in subsequent reports, to the separation of divisions 4, 5 and 6, on the grounds of lack of fossil evidence, and they were placed in a newly erected group of doubtful beds, under the name Cambro-Silurian. The true relations of these beds to underlying rocks of pre-Silurian aspect will be considered further on, when the general structure, as at present understood, is discussed.

Next in order, among the many puzzling and doubtful formations Mascarene discussed in this report, we may briefly glance at the Mascarene series. series.

This group of rocks, so called from its recognized development along the shores of the Mascarene peninsula, on the east side of Passamaquoddy Bay, comprises strata of very different character ; some of which resemble very closely typical Huronian rocks, such as diorites, fine-grained felsites, petrosilex, etc., which are intimately associated with, and, in fact, apparently overlie a series of slates and sandstones, containing fossil shells and remains of plants. Here then is presented a second problem, somewhat similar to that which so long remained unsolved near St. John. The intimate admixture of beds of such great diversity of character rendered the expression of any definite opinion as to the real age of this group exceedingly difficult, and it was accordingly decribed as an isolated series, without being assigned to any definite horizon. A re-examination was made in 1874, and in the report for that year it was stated to be of Upper Silurian age.

The parallelism of structure displayed in this group of rocks is so marked, when contrasted with the earlier apparent superposition of metamorphic Huronian strata on the Devonian east of St. John, that the merits of the question may be here briefly stated.

Bailey and Matthew. Rep. Geol. Sur., 1870-71,

The various strata exposed along this portion of the shore are arranged in five divisions ; of which the lowest consists of hard, grey felspathic slates and olive grey argillites, apparently non-fossiliferous, and having a thickness of about 400 feet.

Div. 2, in ascending order, comprises black and dark grey banded siliceous slates, with obscure plant remains, and with thin bands of conglomerate ; thickness rather more than 600 feet.

Division 3 consists of grey sandy flags and slates with slate conglomerates, in all about 400 feet, the greater part of which contains fossil shells.

Div. 4, principally bright red and green slates, sometimes with a purple tinge, apparently without fossils; thickness 300 feet.

Div. 5, principally hard, red felsites, often porphyritic and of Huronian aspect ; thickness unknown, probably over 300 feet.

L. W. Bailey and G. F. Matthew, Rep. Geol. Sur. 1874-75.

The geological position assigned this series in report 1874-75, as a portion of the Upper Silurian, seems, from the characters of its various divisions, to be somewhat unsatisfactory. Thus, on page 88, it is stated that division 2. or its equivalent on Beaver Harbor, contains plant stems, among which a *cyclopteris*, a *cordaite*, a *sphenopteris*, and the remains of ferns were clearly recognized. The presence of these certainly impart a Devonian aspect to this portion ; while the Upper Silurian fauna in the overlying beds would indicate an overturn or a line of fault at this point. Confirmatory evidence in this direction is afforded by the overlying bands of Huronian looking rocks, which had for a long time been paralleled with the Coldbrook and Kingston groups, and it is very probable that, taking into consideration the crumpled character of the Huronian sediments of Beaver Harbor, with the infolding there of narrow belts of rocks holding Silurian fossils, we may have a similar condition of things in the Mascarene peninsula. Since the date of the report in question, however, no further attempt has been made to work out this problem in a more satisfactory manner.

Ascending in the geological scale, we find the outlines of the Upper Silurian. Upper Silurian considerably changed as compared with all previous reports. The separation of the Kingston group made a great reduction in its aggregate thickness. In Charlotte county, the formation comprised several localities where characteristic fossils were found, such as the vicinity of Oak Bay and the group of islands and promontories lying between Beaver Harbor and the Western Isles. Collections of these were made at various points, notably at Back Bay on Frye's Island and at Oak Bay.

Further east in Kings and Queens counties, the fossiliferous areas east of the granite, and extending to the St. John River, were well defined. These were found to embrace slates, sandstones and conglomerates, with a considerable thickness of hard, quartzose beds and occasionally felsites and diorites, the total thickness being about 5,700 feet.

Along the line between Kings and Queens counties, these rocks were found to overlie unconformably a series of hard felsites and other rocks of Huronian aspect, which form prominent hill features in this area, the debris of which enters into the composition of the basal beds of the Silurian. The age of these strata in the vicinity of Beaver Harbor, as determined by the fossils there obtained, is about that of the Niagara formation.

The Devonian system has, as contrasted with the report of 1865, Devonian. been also largely remodelled. The four divisions of Bloomsbury conglomerate, Dadoxylon sandstone, Cordaite shale and Mispec conglomerate were retained, but the first was reduced by about 2,000 feet of trappean and other volcanic rocks which, from their character, were now referred to the Huronian system, while the Cordaite shales were reduced by the separation of the apparently upper and metamorphic portion, which was erected into the Coastal group as already described.

The total thickness of the four divisions of the Devonian as now existing, was estimated at 7,500 feet, but as the area is profoundly affected by faults, this estimate may not be actually correct, owing to possible repetitions of certain strata. It does not include also the beds of the Perry sandstone group which, upon the authority of Sir

William Dawson, from the evidence of contained plants, had been
assigned to this system, as its upper member, several years before,
and had also, in the report of 1865, been so described by Prof.
Bailey, but subsequently separated on the grounds of want of conformity to the underlying Devonian of Point Lepreau, and from
a supposed resemblance to certain recognized Lower Carboniferous sediments in the vicinity of the St. John River. Various
publications have appeared concerning the fossils of this group,

among which may be mentioned those by Sir William Dawson
in Can. Nat., 1861, on Pre-Carb. flora of Maine, &c.; Flora of Dev.
period, in N. E. America, Geol. Soc., 1862; Observations on Dev.
plants of Maine, Gaspé, &c., Geol. Soc., 1863; Fossil plants of Dev.
and Up. Sil. of Canada, Geol. Sur., 1871. From the evidence, therefore,
of its contained plants, and from its marked resemblance to much of
the Upper Devonian conglomerate of the Gaspé district, it seems
highly probable that this group should also be assigned to the
Devonian rather than to the Carboniferous system. Its distribution
is limited. It occupies the greater part of the peninsula between the
St. Croix River and Passamaquoddy Bay, underlying the town of St.
Andrews, with several adjacent islands, and occurs in patches at the
mouth of the Magaguadavic River and as a considerable area about
the village of Lepreau and on the peninsula which terminates in the
point of that name.

The Carboniferous system was divided into three divisions, the
Lower, Middle and Upper, of which the first, having a thickness of
some thousands of feet, contained the limestones, gypsum, ores of
manganese, and the peculiar mineral albertite, while the Middle and
Upper were briefly described under the general term Coal-measures,
since the examination of this formation had not been carried on
sufficiently to warrant a separation of its various members. The
subdivisions of the Lower Carboniferous were, as stated in previous
pages, accurately given in the first edition of Acadian Geology, when
the bituminous shales and associated conglomerates of the Albert
mine were rightly regarded as the lowest portion. This order of suc-

cession was, however, somewhat modified by Mr. Matthew from subsequent observations made along the line of railway near Norton,
where the bituminous beds had an apparently higher position. More
detailed examination proved that both observers were right, for while
in the sections on the Petitcodiac River, the lowest beds in the formation are, without doubt, the basal conglomerates and Albert shales, at

several points in Kings county the presence of bituminous shales was distinctly observed near the middle of the series. The distribution of the series was, however, given with considerable accuracy, and the presence of a volcanic, intercalated portion in a section along the St. John River in the parish of Hampstead was noted, trappean and felspathic rocks occurring both near its summit and base. This volcanic feature is also well observed at other points, not only along the south-western margin of the formation in Queens, Sunbury and York counties, but in outcrops at intervals in the very centre of the Middle Carboniferous basin about the upper end of Grand Lake. As these rocks, however, had at this time not been particularly studied, many additional facts regarding them were brought to light on later investigations, and have been described in more recent publications.

The opinions stated in this report, 1870–71, regarding the Middle Carboniferous confirmed those expressed by Dr. Robb in 1850, viz : that this formation was of no great thickness and probably represented only by its lower member, as indicated by the presence of the outcrops of metamorphic slate, pointed out by C. R. Matthew, in what was supposed to be the thickest portion of the coal field. The fossils collected were, for the most part, characteristic of the lower horizon, but in several cases appeared to indicate a higher range, embracing Upper Carboniferous forms, which might be derived from unconformably overlying patches of newer rocks or a mingling of the flora of different horizons.

The deposits along the Bay of Fundy, described as of Triassic age, Triassic and Permo Carboniferous. have, as in the case of so many other formations, been long a subject of dispute. Dr. Gesner, in his early reports, included not only the areas mentioned in the report under consideration, as probably of this age, but included also other deposits, some of which, like those of Lepreau and Passamaquoddy Bay, have been seen to be presumably Devonian, while others, farther east, are now known to belong to the Lower and Upper Carboniferous formations. Dr. Robb, on the other hand, held that no rocks newer than Carboniferous existed in New Brunswick.

In many respects, the limited areas referred to in this report do resemble and are undoubtedly the equivalents of the so-called Triassic of Nova Scotia and Prince Edward Island, but as later investigations in these localities render it exceedingly probable that a considerable portion of these rocks, especially along the north side of Minas

Basin, as well as on the island of Prince Edward, really belong to the Permo-Carboniferous, the same conclusions may probably be stated in reference to the supposed similar rocks of New Brunswick. The presence of plant remains in the sandstones at Quaco and Martin's Head, similar in character to those found in Prince Edward Island in what was formerly there regarded as Lower Trias, but now held to belong rather to the Permian, makes it very probable that more detailed study of these rocks will show that they also may be classed as the upper member of the Palæozoic series. In their stratigraphical relations, there is nothing to indicate their higher position, since they rest generally on Lower Carboniferous or Millstone-grit sediments, while the occurrence of undoubted Upper Carboniferous beds in Westmoreland county, similarly placed, and closely resembling these in character, tends to support the present view.

As in Nova Scotia, and to a very limited extent in Prince Edward Island, the stratified portions are penetrated by trappean rocks, which, along the contact have altered the sediments to a certain extent, but never for more than a few feet. The peculiar metamorphosing influence of the Triassic traps can be seen at many points, the sandstones being baked and hardened, sometimes losing their red color and becoming grey, while in places, the beds are thrown into local anticlinals, by the apparent upward thrust of the volcanic matter. The greatest development of these trappean rocks is seen on the Island of Grand Manan, the western half of which is composed largely of basalts, amygdaloids, and trap ashes, with limited exposures of red sandstone, which, near the contact, are impregnated, to some extent, with copper glance, supposed at one time to constitute a deposit of economic value.

There yet remain to be considered two groups mentioned in this report, which, owing to their great diversity of lithological character, have ever been difficult of adjustment in the geological record. These were described under the head of various pre-Carboniferous rocks, and were divided into two portions, styled, from their prevailing colors, the "dark and the pale argillites." The lower members which were generally dark-grey and often flinty in texture, were highly felspathic, and so far as could be ascertained, were devoid of fossils ; the upper while generally of a lighter hue, had a greater preponderance of grey sandstones and shales and in places showed indistinct traces of plant remains. The lower series was found often in direct contact with the great granitic mass of Charlotte and Kings counties, the meta-

Pale and dark argillite series.

morphic action of which was evidenced by the alteration of the clay slates into schists, and by the production of crystals of staurolite, actinolite, mica, &c., in the adjacent beds. No definite fossil evidence could be found to locate this group in the scale of formations, though from the resemblance of its lowest members to the lower portions of the Mascerene series, it was supposed to be not far removed from the horizon of the Upper Silurian. The similarity of much of the upper group to the portion of the Devonian characterized as the Cordaite, together with the presence of plant remains, and its stratigraphical position between the dark argillites and the Lower Carboniferous, caused this portion to be regarded as probably belonging to the Devonian system. Both these series were in the map accompanying the second edition of Acadian Geology, colored as of Lower Silurian age.

Subsequently, in 1875, these pre-Carboniferous rocks, were arranged by Mr. G. F. Matthew, under three heads, viz: Huronian, Upper Silurian and Devonian, with the proviso, however, that some portions, notably of the dark argilites, or lower group, presented strong resemblances to the Lower Silurian of the western part of the province. But later, in the general report accompanying the map of southern New Brunswick, (1878-79,) they were re-arranged, the pale argillites being placed permanently in the Devonian, the doubtful or dark argillites classed as a new group, under the title of Cambro-Silurian, while a large area to the east of and between the granite ridge and the St. John River, in which typical fossils were found at various points, was called Silurian, certain areas of felsites which protruded through these fossiliferous sediments, and formed prominent hills and ridges, being, from their lithological aspect, regarded as true Huronian rocks. *Mr. G. F. Matthew, on pale and dark argillites, Rep. Geol. Sur. 1875.*

Resuming the regular history of New Brunswick geology, the next paper on the subject was by Prof. L. W. Bailey, 1871-72. This dealt principally with the extension eastward of the Huronian system, into Albert and Kings counties, more particularly with reference to the Coastal group, which was found to occupy the greater part of the country bordering on the Bay of Fundy in this direction, and was regarded as constituting the mineral bearing, or copper-belt of this area. Attention was also directed to the presence of a small deposit of supposed true coal measure rocks at Dunsinane, on the railway qetween Sussex and Petitcodiac, (see also Report 1865, p. 157,) *Prof. L. W. Bailey, Rep. Geol. Sur., 1871-72.*

and which was thought might be similar in character to the coal area of Spring Hill, in Nova Scotia, a supposition, however, not verified by later examination of the district. The occurrence of intrusive traps in eastern Albert, of supposed Triassic age, was also noted, and this was considered important, as tending to strengthen the view expressed the previous year, as to the Mesozoic age of certain patches of red sandstone found at points along the coast in the vicinity. The intrusive character of the red granites, both of the great southern area, in Charlotte county, as well as certain masses in Albert and Kings, not only in Silurian sediments, but clearly intruded into older and possibly Laurentian rocks in the south-western portion of the province, was well indicated.

The season of 1872, was devoted to an examination of the central or Grand Lake coal field, with a view to determine more conclusively its economic value by a series of borings, and thus to settle definitely the much vexed question as to the occurrence of workable seams underlying that, which, as a surface seam, had been worked to some extent for many years. No detailed study of this coal field had been attempted since the time of Dr. Robb, whose views on the subject have already been stated. In the report on this area, 1872-73, the entire thickness of the formation above the Lower Carboniferous was estimated to be only 600 feet, which was divided into three portions, of 200 feet each, assigned to the Millstone grit, the Productive Coal measures, and the Upper Carboniferous, respectively. The general thinness of the formation was again conclusively pointed out by the presence of areas of older rocks at various places at the surface, and subsequently by means of the diamond drill, which penetrated the entire thickness of the Middle Carboniferous at the head of Grand Lake, at a depth of not more than 200 feet. Borings carried on during the next three years, at different points, effectually proved the barrenness of the measures and disproved beyond a doubt the erroneous ideas entertained by many concerning the great economic importance of this area.

Bailey and Matthew, Rep. Geol. Sur. 1872-73.

R. W. Ells. Boring operations at Grand Lake, Rep. Geol. Sur. 1872-73.

Later examinations of the Carboniferous basin over its entire extent, embracing an area of 10,000 to 12,000 square miles, have led to a modification of the views then expressed, and it has been satisfactorily established that with the exception of certain small outlying patches of Upper Carboniferous rocks, occurring for the most part along the shore of Northumberland Strait and in the eastern part of Westmoreland county, the rocks of the entire area belonged

to the horizon of the Millstone-grit, and were below the true productive measures of Nova Scotia.

The season of 1873 was devoted to the completion of the geological and topographical map of Queens and Sunbury counties, embracing not only the coal field just described, but the various formations bordering this area on the south, including the hitherto doubtful
groups of the argillite series and the associated Huronian strata.
The conclusions then arrived at, together with the map, were
withheld for several years, pending further investigations, or until
the appearance of the general geological map of southern New
Brunswick in 1878–79, in which, however, the views expressed were
very considerably modified as compared with those held in 1873.

In 1874–75, several papers on the geology of the province were
published. In Charlotte county, these related more particularly to the
Mascarene series as developed about Passamaquoddy Bay, and
in the adjacent part of the State of Maine. From the facts then
observed, the age of the series was inferred to be Upper Silurian, but
as this question has already been fully discussed in previous pages,
nothing further need here be said. The age of the pale argillites
was, from the observations of the preceding season, proved to be,
for the most part, at least, Devonian, while, concerning the lower or
dark argillite group, satisfactory conclusions could not be arrived at.

Other papers also appeared relative to the age, and distribution
of the iron ore deposits of Carleton county, and sundry facts were
also given concerning the structure of the Grand Lake coal field,
derived from boring operations, by which the statements already
made respecting the general barrenness of the measures were fully
confirmed.

In this year also, (1874,) appeared a paper by Dr. Honeyman of
Halifax, Nova Scotia, upon the various formations of the southern
and eastern portions of New Brunswick, in which he paralleled the
Laurentian and Huronian systems about St. John, with the rocks of
the Cobequid series in the adjoining province, and with his lower
Arisaig group of Antigonish. In the northern part of the province,
he regarded the fossiliferous limestones of Dalhousie as probably of
Niagara age, the equivalents of his group, C. Arisaig, and recognized
also the Lower Silurian, now Cambro-Silurian, aspect of the rocks
on the lower Nipisiguit River and Tête à Gauche, in which graptolites were afterwards found of presumably the horizon of the Utica
or Trenton, (see Rep. 1879–80, p. 231,) while the later age of the

Bailey and
Matthew,
Charlotte
county. Rep.
Geol. Sur.
1874-75.

R. W. Ells.
Iron ore of
Carleton
County, Rep.
Geol. Sur.
1874-75.

R. W. Ells.
Boring
operations,
Rep. Geol. Sur.
1874-75.

Dr. Honeyman,
Trans. N.S.
Nat-Sci. Inst.
1874.

intrusive traps in the Dalhousie section, first pointed out by the late Sir Wm. Logan, (Rep. Geol. Sur., 1884,) was proved by the alteration of the contiguous beds and by the presence of Upper Silurian fossils contained in the trappean mass itself.

Bailey,
Matthew. Eells
The pre-
Carboniferous
of southern
New
Brunswick.
Rep. Geol. Sur.
1875-76.

In 1875–76, a report on the pre-Carboniferous formations of Southern Queens and Sunbury counties, with various rocks in northern Kings, was presented, illustrative largely of the work done in 1873. These were now divided into three groups, the crystalline or older felsitic series, Coldbrook or Huronian, the dark argillites which were now considered to be Upper Silurian, though possibly containing areas of Lower Silurian, and the pale argillites, which were again asserted to be Devonian, from the evidence of the contained plant remains. The Upper Silurian, however, was now made to contain a considerable area of what was in 1870-71, regarded, on lithological grounds, as pre-Silurian or Huronian.

Mr. G. F.
Matthew,
Charlotte
county.
Metamorphic
rocks.
Rep. Geol. Sur.
1876-77.

In 1876-77, an interesting report by Mr. G. F. Matthew, on the geology of Charlotte county, was published. In this, the aspect of certain supposed pre-Silurian rocks in the south-western portion, was discussed, more especially in their development west of the lower part of the Digdequash River. These, for the most part, highly crystalline rocks, consisted of diorites, fine and coarse, hard schists, porphyritic and slaty felsites and gneissoid sandstone, which often present the appearance of true fine-grained gneiss. They form an area of considerable extent, and are intersected by intrusive granites which may be spurs from the main granitic mass of that district. They cross Oak Bay and extend into Maine, and also form a conspicuous band on the St. Croix River, in the vicinity of, and for some distance above, St. Stephen and Calais. From the fact that these older volcanic rocks do not penetrate recognized Silurian strata in that vicinity, as well as from their marked lithological resemblance to the pre-Cambrian of St. John county, they were at that time regarded as also of Huronian or Laurentian age.

In this report, also, the dark argillites which have a considerable development in this county, forming an extensive belt which stretches diagonally across from the St. Croix River to its north-east corner, were considered as of Upper Silurian age, the rocks of which system were arranged in five divisions, the three lowest being paralleled with the divisions of that system, as stated in the report for the preceding year, whilst Nos. 4 and 5 of that series were supposed to represent Nos. 3 and 4 of the Kingston group. The latter was, by this arrangement, removed from the position assigned it in 1870-71, as a portion of the Huronian to the upper part of the Upper Silurian, chiefly from its supposed superposition on the St. John group, as inferred from the presence of pebbles of black slate in the conglomerates near its base, which were held to be derived from beds of Primordial age.

The crystalline and Huronian aspect of the Kingston group, as a whole, can be well seen by reference to the sections given on the New River ; the character of the several divisions there seen being as follows :—

• Green chloritic and granitoid rocks.

Dark, porphyritic, slaty felsite, with grains of clear quartz.

Grey clay-slates and diorites.

Chloritic and felspathic slates and grits, with slate conglomerate overlaid by about 11,000 feet, principally schists both micaceous and hornblendic, the latter predominating, crystalline felsites, and capped by chloritic and felspathic gneiss.

No fossils were found in any of the strata, but it was distinctly stated that this group in Charlotte county constituted the mineral belt in the same way as the Coastal group to the eastward, and which was also formerly considered to occupy nearly the same relative position as now assigned to the rocks under discussion. As, however, uncertainty still existed as to the true position of the Kingston group, the publication of the map and section accompanying this report was, for the time, deferred.

In the report for 1876–77, the relations of the great mass of red syenite and granite, which form so important a feature in the geology of this county, to the associated rocks, were also carefully given. That they were clearly newer than the overlying Upper Silurian was evidenced by the metamorphism of the slates in contact, shown by the production of various crystals, as also by the presence of numerous faults and dislocations, and by the penetration of contiguous strata by numerous dykes and veins of granite, which, proceeding from the main mass, cut the adjoining beds with sharply defined lines of contact.

The same volume contained also a report on the Lower Carbonifer-ous of Albert and part of Westmoreland counties, by Prof. L. W. Bailey and the author of this paper. This had particular reference to the distribution and economic value of the portion known as the Albert Bailey and Ells. on Lower Carboniferous of Albert County. Rep. Geol. Sur., 1876-77.

or bituminous shales, of special interest, from their containing the remarkable mineral, Albertite. In this paper, also, the pre-Cambrian aspect of the rocks of the Caledonia Mountain range was pointed out. The divisions of the Lower Carboniferous there given were five in number, representing an exposed thickness of 4,150 feet, separable into two unconformable series by a break between the bituminous shales and associated conglomerates of the Albert Mines, and the sandstones and conglomerates of division 3, while the great masses of gypsum and limestone were found to lie near the upper part of the formation.

The season of 1877-78 was devoted to further detailed work, prin-cipally bearing on the relative positions of the various divisions of pre Silurian rocks in Kings, St. John, and Albert counties, the results of which were stated in three papers by the writer, Prof. Bailey and Mr. G. F. Matthew respectively, with a supplementary report by the latter on the superficial geology of the southern part of the province.

R. W. Ells. Pre-Cambrian of southern New Brunswick. Rep. Geol. Sur. 1877-78.

Anticlinal structures pointed out.

In the first of these, the structure of the pre-Cambrian ridges was given as follows :—three main anticlinals, situated to the south of of Kennebecasis Bay, extend from the vicinity of St. John north-easterly, and roughly parallel to the north shore of the Bay of Fundy. In Albert county, and for some distance to the west, the two more southerly anticlinals are apparently overturned ; the con-tained synclinal occupying the area along the Shepody road, and for some distance on either side. The most southerly anticlinal forms the crest of the mountáin ridge in rear of Hopewell, whence, extending westerly, it crosses the Upper Salmon River, some three miles below the Shepody road, and comes to the coast a short distance west of Martin's Head. On the Albert county line, it has a breadth of three to four miles, and is distinguished by the presence of syenite and gneiss, often greenish or protogine, with schists, felsites and dolomites.

The second anticlinal extends from the eastern part of the Cale-donia Mountain range in the· vicinity of the Albert mines, southerly through the southern part of the Mechanic Settlement and reaches the Shepody road a short distance west of the Kings county line, on which road it can be traced for nearly eight miles. Further west, it can be seen on the Big Salmon River, about four miles from its mouth, beyond which it is apparently concealed by overlying sedi-ments. This anticlinal, in Albert county, is flanked at one point by crystalline limestones, of Laurentian aspect, similar in character to those in the vicinity of St. John.

The third anticlinal comprises the syenite, gneiss and limestone of St. John and vicinity. Westward, it can be readily traced to its termination on the coast at Lepreau Harbour, although in places, concealed by Lower Carboniferous and Devonian strata, while eastward it extends along the south side of Kennebecasis Bay, appearing also in the islands and headlands of the southern extremity of the Kingston Peninsula, uninterruptedly to a point south of Hampton station, from which point it is about four miles distant. It re-appears from beneath Lower Carboniferous beds, four miles further east, and forms a narrow ridge, eight miles in length, when it again becomes overlapped by the . Carboniferous formation, and does not again re-appear, but on the Kings county line an anticlinal, in the continuation of the one just described, brings up the Lower Carboniferous beds from beneath the Millstone grit, and probably indicates the extension of this third axis, which, in this case is parallel with the second anticlinal of the Caledonia Mountain.

A series of chloritic, talcose and felspathic schists, ash rocks, purple grits, and conglomerates occupy the first synclinal in Albert county. These are apparently unconformable to the rocks of the underlying anticlinals just described as well as to their supposed •equivalents in Kings and St. John counties. They occupy a considerable area along the Shepody road, and the greater part of the coast of the Bay of Fundy, between Point Wolf and Melvin's Beach, about five miles east of Quaco.

Along the northern side of the second anticlinal in Kings county, these rocks, which constitute the Coastal portion of the Huronian, are to a great extent apparently replaced by a considerable thickness of brecciated, siliceous and felspathic, ash rocks and diorites, which, according to Prof. Bailey (see his report on this group, 1877), compose the greater part of the pre-Cambrian series to the south of the main Laurentian axis or anticlinal No. 3. These are regarded by him as older than the Coastal rocks which occupy the synclinal just mentioned and which form the shore series, and are described in ascending order under two heads, the first or felsite petrosilex group consisting largely of the following rocks :— Prof. L. W. Bailey on pre-Cambrian of southern New Brunswick, Rep. Geol. Sur. 1877-78. Coastal rocks.

Red and grey felsites, blue, grey, reddish and black petrosilex and breccias, diorites, and amygdaloidal ash rocks and ashy conglomerates.

Grey felspathic sandstones and conglomerates, the whole representing the Coldbrook of former reports and now regarded as the oldest member of the Huronian.

The second or upper division consists of the following rocks :—
Chloritic schists, greenish, grey and purple ash rocks and amygdaloids, with purple conglomerates.

Pale grey pyritous and rusty-weathering felsites and felspathic quartzites.¹

Hydromica schists, chloritic and felspathic schists, grey clay-slates and purple conglomerates, with beds of hæmatite, styled Coastal in former reports, and now regarded by him as the upper member of the Huronian.

Mr. G. F. Matthew. Pre-Cambrian of southern New Brunswick. Rep. Geol. Sur. 1877-78.

The third report, by Mr. G. F. Matthew is confined to the Kingston group and principally to its development in the Kingston peninsula, where from the discovery of fossiliferous ashy rocks and slates of Upper Silurian age along the south side of the Long Reach of the St. John River which, at several points on the shore, have a dip inland, it was inferred that a large portion of the rocks of this group were stratigraphically above the fossiliferous beds and consequently newer.

Kingston group.

He, however, at the same time, in tabular form, paralleled the Kingston of this locality and its extension westward into Charlotte county, under the heading Upper Silurian, with the recognized Huronian of St. John, in which the lithological resemblance of the two groups is such as to strike the most casual observer.

The Huronian character of the crystalline and metamorphic rocks of southern New Brunswick, was next pointed out by Dr. T. Sterry Hunt, in his report to the Geological Survey of Pennsylvania, published in

Dr. T. Sterry Hunt, 1878.

1878, and subsequently in a paper read before the American Association for the Advancement of Science, (1879), in which their unconformity to the Laurentian of the St. John area was indicated as well as their lack of conformity to the overlying primordial slates. In the "supplement to Acadian Geology, Sir Wm. Dawson, 1878, several

Bailey, Matthew and Ells. Final Report on southern New Brunswick, 1878-79.

allusions to this province also were made, the principal points of which have been already considered." In this year also, the final report on the geology of southern New Brunswick appeared, accompanying the map of that area, which had been so long deferred. From a careful consideration of all the data in the possession of its several authors, which had been accumulating for years, the various geological lines were there laid down, and the most recent and probable views concerning the horizons of the different formations stated. As no subsequent publication on this portion of the province has appeared, we will briefly state the position and arrangement of the geological systems as there expressed, with some probable changes

which may hereafter be made in certain groups, about which sufficient was not at that time known to pronounce definitely as to their
exact age and relations.

It must be premised that in speaking of the geology of southern
New Brunswick, we include that portion only which lies to the south
of the great central Carboniferous area, in so far as relates to the
pre-Carboniferous rocks, and in regard to the central area, that lying
to the south of the parallel of 46°, which passes a short distance
north of Fredericton. The study of the northern and south-eastern
portions having been taken up at a date subsequent to 1878, these
areas will be considered further on.

The geological systems at present recognized in that portion of the
province now under consideration, are as follows :—

Laurentian.............Lower.
                 Upper.
Huronian......... ....Formerly divided into Coldbrook, Coastal
                 and Kingston.
Cambrian............ Primordial Silurian Acadian,or St. John group.
Cambro-Silurian......Formerly Middle Silurian.
Silurian.
Devonian.
Carboniferous........Lower.
                 Middle or Millstone-grit.
                 Upper of eastern Westmorland county.
Triassic.
Intrusive rocks.......Granites, syenites, diorites, diabases. Trappean
                 rocks of Carboniferous and Triassic age.
Post Pliocene.

In the general geological map of the southern part of the province, Laurentian.
the area colored as Laurentian embraces principally that portion
which, in 1860, was separated from the Devonian, under the title of
the Portland group. It has, however, since then, been considerably
extended ; and west of St. John City, it embraces the greater part of
the county of St. John, extending across into Charlotte county, and
terminates westward, at Ragged Head, which forms the south-western
point of Lepreau Harbor. Though apparently divisible into two
unconformable series, it was not deemed advisable to color these separately, but the limestones were designated, wherever known to
occur, by a distinct tint, and thus the upper member was, to a certain

3

extent, made conspicuous. The anticlinal structure of the lower member is well marked, and its probable extension north-easterly has already been given. The limestones in the vicinity of the St. John River flank the axis on either side. These are well developed along the south side of Kennebecasis Bay, and from South Bay they extend about four miles west of the river, where they apparently form a basin. Eastward they appear as outliers on the syenitic rocks beyond Rothesay, and at several points as far east as Norton. West of St. John, they extend entirely across the peninsula which divides Pisarinco from Musquash Harbor on the south flank of the main anticlinal and on Point Lepreau are also exposed in synclinal form, resting upon greenish gneiss and granitic rocks. The width of the Laurentian area, comprising both divisions, is from six to seven miles.

In report 1870-71, it was stated that a belt of gneissic rocks, bearing much resemblance to those just described, extended from the head of the Long Reach of the St. John River, along the north side of that sheet of water, to L'Etang Harbor, forming an anticlinal ridge flanked by Huronian sediments. On Frye's Island, near the southwestern extremity of this axis, the syenitic portion is overlaid by crystalline limestone, in character precisely similar to those which occur with Laurentian rocks elsewhere, and the whole belt was considered as probably a parallel ridge, contemporaneous with the belt just described. While, however, this view was held concerning the age of these rocks, the difficulty of separating them precisely from the overlying Kingston group, owing in large part to the wooded character of the country which they traverse, was such that it was thought best to include this series in the general pre-Cambrian color without definitely specifying its exact position. Lying off the coast in this direction, the group of islands called The Wolves, were found to be the extension of the main Laurentian anticlinal, and to be composed of the greenish syenites and other rocks of the lower series in which, however, no limestones appeared.

Further west in the parishes of St. Patrick, St. Croix and St. Stephen, syenitic, dioritic and gneissic rocks are common. These in the vicinity of the St. Croix River, present characters very similar to those of the fundamental rocks of the St. John area, and were at one time described as of the same horizon. They are intersected by large dykes and masses of intrusive syenite, some of which are of much later age than the rocks with which they are associated. In this locality also, the authors found much difficulty in separating the

the newer intrusive rocks from the older portions, and all these areas of syenites, &c., of whatever age, were therefore included in the same color, though the outlines of the newer are indicated on the map in a general way.

It must therefore be fully understood that in the area indicated as granite west of the Digdeguash River, and for a mile or so to the east, a considerable portion may possibly be of pre-Cambrian age."

In Albert county, also, attention was called in 1870-71, to the presence of rocks similar to those of the typical Laurentian of St. John, and the supposition was then advanced that these occurred as ridges surrounded by Huronian strata. This view, in so far as the stratigraphical relations of the several groups were concerned, was confirmed in 1878, and the probable Laurentian age was indicated by the presence of crystalline limestones and dolomites. The older portions of these ridges, as we proceed westward into Kings county, speedily become concealed by the unconformable overlap of the Coldbrook and Coastal groups, from which, in lithological character, they are markedly different. They contain also a considerable area of coarsely crystalline diorites, holding magnetite, which in places be comes, almost a pure hornblende rock. *Supposed Laurentian of Albert county.*

A similar area of magnetic diorites is found in northern Kings county near the Scotch settlement. This was referred to in 1870-71, and supposed to be of Laurentian age, as well as several areas west of the St. John River, in the valley of the Nerepis stream. Dioritic rocks occur in Charlotte county, near St. Stephen, which are also probably of this age. They contain serpentine and chrome iron in small quantity and underlie, unconformably, hard micaceous quartzites. East of St. John also, near Dolin's Lake, bands of hard crystalline felspathic, sometimes dioritic rock occur, associated with hypersthene and magnetic iron which have been considered by Dr. T. S. Hunt, as identical with some varieties of Norite rock of the Upper Laurentian or Labradorian series. They are associated with gneisses, also presumably of the upper division. *Magnetic diorites.*

The limestones near St. John, are in places highly serpentinous and have furnished indications of the presence of Eozöon. At Pisa- rinco, however, serpentine is seen to cut talcose and chloritic rocks, associated with limestones, as a true dyke six feet wide. At the falls or the St. John River, also, deposits of plumbago of large extent are found in the slates and limestones. It is worthy of remark that in *Serpentinous limestone of St. John with Eozöon.*

all the supposed Laurentian areas of Charlotte county, no traces of limestone have as yet been found, except at Frye's Island and Lepreau.

Huronian. The history of the Huronian system, with the many changes which have taken place in relation to its various groups, has already been somewhat fully stated and we will, therefore, here present only the views now held regarding its distribution and mineral character.

The term Huronian does not appear on the general map, for the reason that, owing on the uncertainty as to the age of the several groups, some of which have just been described, it was thought best, until their true relations to the Laurentian or Huronian systems proper could be accurately worked out, to include these under a general term, pre-Cambrian, since, in position, they were found to unconformably underlie the fossiliferous Cambrian, at several points.

Discussion of the several groups of Coldbrook, Coastal and Kingston. The former division of the Huronian into Coldbrook, Coastal and Kingston was exceedingly unsatisfactory, inasmuch as, while they were all regarded as portions of the same system, their respective horizons could not with certainty be determined. Thus in 1870-71 the Coldbrook is the first described after the Laurentian, leading to the inference that this group comprised the lowest members of the Huronian system, while at the same time its upper part was held to represent the basal beds of the St. John group, to the exclusion of the Coastal and Kingston. In the last report by Prof. Bailey, the Huronian south of the Kennebecasis is divided into two groups only, the Coldbrook and Coastal, of which the latter is held to be the newer. The older is distinguished by the term "felsite petrosilex group," from the prevailing character of the rocks which compose it, while the newer is termed the "micaceous and chloritic or "schistose group," for the same reason. It is evident, from a consideration of the rocks of the Coldbrook, that they are for the most part of volcanic origin, the prevalence of breccias and brecciated agglomerates, diorites and ashy rocks, which make up the bulk of the group, presenting a marked contrast to the more slaty and schistose rocks of the Coastal, though the latter has abundant evidence also of volcanic action in the presence of ashy and agglomerate rocks; it has also a considerable thickness of clay-slates, with purple grits and conglomerates. It is exceedingly doubtful, however, whether a separation such as has just been proposed, into an upper and a lower division, can be successfully made on stratigraphical grounds, owing to the difficulty of determining which group rests primarily on the

Laurentian, since in some places this position is held by the Coldbrook, while at others the Coastal is in direct contact.   This difficulty becomes more apparent, when we consider that the original
Coldbrook of the earlier reports, so named from the place where
first studied, is now regarded as Coastal.   It is possible, that both
divisions may be contemporaneous, or that the Coldbrook or volcanic
portion may be the more recent, and this latter view receives a
certain amount of support from the stratigraphical evidence seen in
the overturned synclinal of Albert county, the rocks of which rest
upon the basal ridges and are undoubtedly of the Coastal or Schistose
type, (see Rep. 1877–78).   These occupy the southern portions
of eastern Kings and St. Johns counties, and, while in their western
extension they overlap the ridges just mentioned, they are in turn
apparently overlaid by the volcanic rocks of the Coldbrook group,
the Coastal not appearing at all in the northern portion of the main
pre-Cambrian area south of the Kennebecasis, unless we except
certain doubtful and limited areas lying to the north and east of the
Loch Lomond lakes.

There yet remains to be considered the third division of Huronian
rocks, viz: the Kingston group, which, also receiving its name from Kingston.
the place where first studied, the peninsula lying between the Long
Reach and Kennebecasis Bay, has also a considerable development
westward in Charlotte county.   The various changes of opinion
respecting the position of this group have already been stated, and
we will, therefore, give the reasons, lithological as well as stratigraphical, which seem to us conclusively to prove that these rocks
should be considered an integral portion of the Huronian system,
rather than to belong to a later horizon.   The character of the rocks
which compose this group are well given in the report of Mr.
Matthew, already referred to, for the Kingston area and for its extension westward, and are here presented for the sake of comparison :—

Compact, dark-grey, diorite and fine-grained, flesh-red felsite.

Hornblende and mica-schists, schistose diorite, and dark felspathic
slates, with grey argillites.

Fine-grained mica-schist, silico-felspathic gneiss and schistose
felsite.   Chloritic gneiss and syenite, with thin beds of limestone,
argillites, &c.

In addition, beds of slate conglomerate occur, and chlorite and
epidote are found in veins of considerable size.

That the Kingston group is entirely distinct from and older than the Silurian was pointed out in 1870-71; but the pre-Cambrian aspect is more clearly indicated, if we glance at its structure, as seen along the St. John River. By reference to the general report, 1878-79, it will be seen that the pre-Cambrian rocks of southern New Brunswick, are arranged in a series of approximately parallel anti-clinals, the synclinals being occupied by fossiliferous Cambrian strata.

These anticlinals were arranged in fine divisions, of which Nos. 1 and 2 were considered of Laurentian age. No. 3 Coldbrook, No. 4, Coastal, and No. 5, Kingston.

This arrangement does not, however, necessarily indicate the true stratigraphical position of the various groups ; for, while we have seen that the Coastal has not yet been definitely proved to be the upper portion of the Coldbrook, it is evident that the Kingston includes, in part, at least, rocks common to both divisions 3 and 4, and possibly even lower.

The extension of the anticlinals 1, 2 and 3, has already been given. Of the latter two, which pertain to the Kingston Peninsula, No. 4, beginning at Clifton, on the south side, extends north-east, past the head of Kingston Creek, towards Dickie Mountain, while No. 5, commencing near the western extremity, and well seen at Milkish Creek, keeps along the north side of the peninsula, and crossing the entrance of Belleisle Bay, is probably continued along the north side of that sheet of water into the Scotch settlements.

A sixth ridge bounds the north side of the Long Reach and forms the eastern extremity of that which, in Charlotte county, was stated to underlie Huronian sediments, and to be possibly of Laurentian age.

That all these ridges are, at least, pre-Cambrian, is plainly indicated by the presence of unconformably overlying areas of Cambrian slates, which occur in the form of basins of greater or less extent, the limits of which have been largely affected by denudation.

Primordial
rocks.

The most northerly of these, definitely recognized by fossils, (though it must be remembered that in the Report of 1870-71, other areas still further inland of Primordial age, but in which no fossils have yet been observed, were supposed to exist,) is found on the west flank of the sixth ridge; but along its south side, a much more important belt occurs, which occupies the greater part of the north shore of the Long Reach from the mouth of the Nerepis to Jones Creek, beyond which, it is largely concealed by Silurian beds. It also appears in several small islands in the river and on the point at the mouth of

Belleisle Bay, where it overlies the more northerly of the two King-ston anticlinals. That this basin of Cambrian rocks occupies the entire valley of the St. John River at this place is highly probable ; although its contact with the Kingston group along the south side of the Reach is partially concealed by fossiliferous strata, which are possibly the prolongation southward of the Silurian area of Jones Creek, since the overlap of the Primordial on the Kingston rocks is well seen at both extremities of the basin. It is probable also that the horizons of the two ridges of metamorphic rocks which extend on either side of the Long Reach, is not very different ; although westward of the St. John River, the Kingston group proper appears to occupy a synclinal between the true Laurentian of the Lepreau axis and the older ridge north of L'Etang. This position would also correspond, on stratigraphical grounds, with that assigned to it from lithological characters as largely representing Divisions 3 and 4 of the pre-Cambrian scale, but in this part of Charlotte county, the Pri-mordial has not yet been definitely recognized.

The presence of slaty conglomerates in the rocks of the Kingston group, the pebbles of which were, at one time, regarded as derived from Cambrian slates, cannot be held as conclusive evidence of the later age of these crystalline rocks ; since the fragments may, with equal reason, be supposed to be derived from the bands of black and graphitic slates, which, at many points, form an integral portion of the Laurentian area, as already stated, and unconformably underlie the Huronian.

In the areas now colored Cambro-Silurian, extending along the county line of Kings and Queens, belts of rock which very closely resemble those just described, are found. They consist of hard fel-sites, often porphyritic, with felsitic, chloritic and talcose schist, and fine and coarse diorites. After many ineffectual attempts to separate these satisfactorily from other slaty beds, with which they are intim-ately associated, but which are, without doubt, of much later age, the greater part of these areas was included, provisionally, in the meta-morphic Cambro-Silurian system. These areas of Huronian-looking rock are in no place of great extent, but appear as crests of ridges exposed by denudation and protruding through the overlying strata.

In fact, the greater part of the formations of southern New Bruns-wick have been so intricately folded, that the problem of deciphering their exact age has been a very difficult one ; and it is to be presumed that with the accession of new light, derived from the study of similar rocks elsewhere, other important changes will be found necessary.

Islands in Passamaquoddy Bay. The views concerning the geology of the islands in Passamaquoddy Bay, called the Western Isles, have undergone the same changes as already described in connection with the pre-Cambrian rocks farther east. The group of The Wolves, consisting almost entirely of syenitic and gneissoid rocks of Laurentian aspect, are, as intimated, the western prolongation of the Lower Laurentian axis from the mainland at Lepreau, while Campobello and Deer Islands, together with the greater number of the smaller islands lying between these and the shore, represent, probably, the extension of the Kingston group. Ores of copper and iron are common, in which respect the rocks resemble those of the metamorphic Coastal belt east of St. John, as well as in their general lithological character, while the eastern portion of the great island of Grand Manan, probably marks the westward extension of the Huronian area lying to the south of the Laurentian axis, which extends through The Wolves.

Similar rocks of England and Wales. The resemblance of the pre-Cambrian rocks just described, to those of certain areas in England and Wales, regarding the age of which much controversy has lately arisen, is very marked. These, by Dr. Hicks, Prof. Bonney and others, are held to be of pre-Cambrian age. They are also divided into three groups, the Dimetian, Arvonian and Pebidian; of which the lower or Dimetian corresponds in lithological characters with the Coastal of Kings and Albert counties, the middle is very similar to the lower Coldbrook, while the tuffs and slates of the Pebidian, much resemble the upper part of that group. By Prof. Geikie, however, none of these rocks are regarded as older than Cambrian; the Pebidian being held by him to form the lowest member of that system, while the other two comprise sundry masses of granite and kindred rocks intruded among the Cambrian strata.

Primordial ocks. The rocks of the St. John group, or Cambrian, were, until 1865, as already stated, regarded as an integral part of the Devonian system. The early researches of Messrs. Hartt and Matthew on these rocks have been alluded to in former pages, but of late years their fauna has been particularly studied by the latter gentleman, whose investigations have brought to light many interesting facts relative to the palæontology of the group, the results of which have appeared, from time to time, in several papers contributed to the Royal Society of Canada and elsewhere. In character, the rocks may be said to consist of purple sandstones and conglomerates, with shales, the former apparently derived from the underlying Coastal and Coldbrook groups, and which constitute the basal portion of the formation. These are succeeded by red and greenish-grey argillites, often micaceous, whitish and grey sandstones, with grey and dark grey sandstones and shales, the latter of which are often fossiliferous, and contain a variety of trilobites, such as *Conocephalites, Microdiscus, Agnostus* and *Paradoxides*, all of Primordial type, together with *Lingula, Obolella, Discina* and *Orthis*. They lie in well defined basins upon the Huronian and Laurentian rocks, of which six, at least, are known. Among the most important are the areas on the St. John River, already described, and the great development which extends from the city of St. John, northeasterly, past the Loch Lomond lakes, towards the upper part of the Hammond River; the other areas consisting of outlying patches, which have apparently escaped denudation, and which serve to illustrate the great dislocations by which, in many cases, these rocks have been affected. The smaller areas about the Kennebecasis Bay are overlapped by Lower Carboniferous sediments.

Other areas, which were supposed to be of this age, have been already noticed in preceding pages, but, though no fossils have been found in these to indicate their exact horizon, their position upon beds which are identical in character with recognized Huronian elsewhere, renders it highly probable that they may belong to the Primordial zone. Until, however, such fossil evidence has been obtained, or their true position otherwise established, these detached areas are, for the present, included in the Cambro-Silurian system.

This system, as it present understood, comprises all the formations Cambro-Silurian. between the Chazy and the Hudson River or Lorraine shales, both inclusive. In New Brunswick, the subdivision into intermediate groups has not yet been attempted. The system comprises certain rocks known, for the most part at least, to underlie the fossiliferous Silurian, and to overlie the Cambrian. The characters of these have already been given.

. In their distribution, the rocks of this system may be said to form a belt, principally of metamorphosed sediments, extending from the western boundary of the province, diagonally across the county of Charlotte, to its north-east corner, flanking on its north side the great granitic area for the greater part of the distance. Thence it extends to the River St. John in the vicinity of Hampstead, and, reappearing on the eastern bank continues along the county line of Kings and Queens for a further distance of twenty miles. The northern limit is fixed, for the most part, by the unconformably overlying interior basin of Devonian sediments; but to the west, at the head of Oak Bay, and to the east of the granite area in south-western Queens, it is overlapped by strata of Silurian age. Much of the country occupied by these rocks is exceedingly difficult of access, and it is presumed that a more detailed examination, when practicable, will change somewhat the outlines of the area as at present defined.

Silurian.   The rocks of the Silurian system are not extensively distri-buted in the southern part of the province. Along the coast of Charlotte they are found for the most part in small lenticular basins, which have been infolded with areas of pre-Cambrian or other rocks along the shores of Passamaquoddy Bay; the association being intricate and often obscure, as in the case of the Mascarene series. Certain beds are, however, well defined by the presence of charac-teristic fossils, which are probably of the horizon of the Niagara formation. The fossiliferous beds are also well seen around the upper part of Oak Bay, where they rest upon the Cambro-Silurian just described, and from this place they extend in a continuous belt along the south side of the granite area as far east as the New River, a distance of thirty miles. In the southeastern part of the province, they have not been definitely recognized, but along the St. John River, in the vicinity of the Kings and Queens county line, they form a considerable area, abounding in fossils at many points, and resting unconformably upon rocks of the Huronian, Canbrian and Cambro-Silurian systems. An interesting band occurs along the Long Reach, resting upon Cambrian grits and congomerates, and on the east side forming a narrow strip of ashy rocks, which resemble, in some respects, beds of Coastal or Coldbrook age. They are, how-ever, clearly distinguished by characteristic fossils, both corals and shells. They dip inland at a moderate angle, as if underlying the crystalline rocks of the peninsula, but this position is only an appar-ent one, since, a short distance from the shore, the contact between the two sets of beds is well seen, the old Kingston felsites and gneissic schists dipping towards the river at a high angle; against these the Silurian rests unconformably. The somewhat obscure stratigraphy at this point has, in the past, led to much misunder-standing as to the proper position of the Kingston rocks, but the true structure in this section is undoubtedly a synclinal in the pre-Cambrian overlaid by Cambrian fossiliferous rocks, which, in turn, are partially overlapped by Silurian strata.

Devonian   The formations which compose the Devonian are of particular interest, from their intricate entanglement with others of much greater age; which, while really underlying them at a great distance in the geological scale, are apparently superimposed upon the fossiliferous rocks of the system ; an order of things which, in the early history of its geology, as already shown, naturally led to much confusion as to the true relations of the various groups. This system is also interesting from its wonderfully rich flora and fauna in the vicinity of St. John and Lepreau ; more especially from its yielding the earliest remains of insect life yet known, at least in America. The flora of the system both from New Brunswick and Gaspé has been carefully studied by Sir William Dawson, whose labors in this direction have appeared in a series of papers ranging in date from 1859 to 1882, and have been readily accepted as standards for the determination of its various formations.

The areas of Devonian along the coast are somewhat limited ; the principal, occurring to the east of St. John Harbor, where the rocks form a double synclinal which extends inland for about eight miles, and occupies the shore from the mouth of Little River, nearly to Cape Spencer. West of the harbor, these rocks occupy a narrow strip along the beach, near Carleton, and constitute the " Fern Ledges " in this vicinity.

A second area extends between Musquash and Lepreau harbors, resting upon Laurentian rocks in a series of folds, of which at least three can be clearly recognized.

These areas, at Lepreau Point, are in turn unconformably overlapped by sandstones and conglomerates, the probable equivalents of the Perry Sandstone group of Maine, the aspect of whose contained plants led to their being regarded by Sir William Dawson as constituting the upper member of the Devonian, but which were afterwards believed by Prof. Bailey to belong rather to the Lower Carboniferous, principally on lithological grounds, but also from a lack of conformity to the underlying Devonian sandstones and shales of this place. We may also include in this system sundry strata found at Wright's Head, on Beaver Harbor, in which are found remains of ferns and other plants, described in the section on the Mascarene series.

But by far the largest area is found underlying the central Carboniferous basin. This, beginning at the boundary of Maine, extends almost to the St. John River, having in Charlotte county a breadth of about twelve miles and bounded on either side by rocks of Cambro-Silurian age. It doubtless underlies a very considerable portion of the Carboniferous area of central New Brunswick, since at several places as at Coal Creek, near the head of Grand Lake, and on the Canaan River and its branches, strata of Devonian aspect come to the surface in the beds of the streams and in low cliffs with cappings of Millstone-grit. These Devonian strata have apparently participated in the series of disturbances which have affected all the formations of the southern part of the province older than the Carboniferous.

Carboniferous formations. The general characters of the Carboniferous formations, as given from the time of Dr. Gesner, down to the publication of the general map in 1878–79, have been given in considerable detail in the pre ceding pages.

Since that date, however, a large amount of work has been done on the rocks of this system in the northern, eastern and central por tions of the province, the results of which, more particularly in refer ence to the Middle and Upper formations, will be briefly stated.

The total area of Middle Carboniferous rocks is about 12,000 square miles, throughout which great uniformity of character is apparent. A thin seam of coal extends over the greater portion, which crops out, not only at the places where worked for many years at Grand Lake and on the Richibucto River, but at various other points along its entire border. This seam, while varying slightly in thickness, in no place, however, amounting to more than 20 or 23 inches, is very uniform in general character ; and it is probable that all the outcrops noted pertain to the same seam, brought to the surface at intervals by a series of low anticlinals which affect the measures throughout their entire extent, following generally a course parallel to the pre Cambrian ridges which surround the basin on the south and north-west. Throughout the counties of Kent, Northumberland and Gloucester, which comprise the area bordering on the Gulf of St. Lawrence, the character of the sediments is similar to that of the recognized Millstone-grit areas of the Bay of Fundy coast, though the horizontality of the beds is such that but little information can be obtained from surface examination as to the entire thickness of the formation. In eastern Westmorland, the investigations of 1884 resulted in the discovery of many interesting facts in regard to the structure and thickness of the several formations comprised in this system. The Lower Carboniferous beds in the vicinity of the Petitcodiac River have a thickness of 4,100 feet ; but further east, along the coast of Cape Maringouin, near the head of the Bay o

Fundy, where fine sections of these rocks are exposed, an additional thickness of not far from 2,000 feet of red marls and sandstones is seen to overlie the gypsiferous portion which constitutes the upper member of the formation. This gives a total thickness of over 6,000 feet or very nearly that found in northern Cumberland county, Nova Scotia.

The thickness of the Middle Carboniferous in the central basin, as given in the report for 1872–73, has also been considerably changed by the study of the rocks further east. Thus the three divisions above the Lower Carboniferous were at that time regarded as follows :

> Barren measures probably Millstone-grit.......... 200 feet.
> Productive.................................... 200 "
> Upper Carboniferous .. .. 200 "

This view was modified in 1878, and the whole interior area was referred to the horizon of the Millstone-grit. The sections of this formation in the vicinity of Dorchester, have shewn that it has a thickness of at least 1,000 feet, and that it is directly over-lapped by the Upper Carboniferous formation which extends thence over almost the entire eastern part of Westmorland county, occa-ionally broken by the presence of ridges of Millstone-grit rocks. Where the contact is not with the Millstone-grit, the Upper forma-tion rests directly on the Lower Carboniferous, which arrangement is also seen in the adjoining province, in the area lying south of North-umberland Strait.

The rocks above the Lower Carboniferous area are as a rule com-paratively undisturbed, but at points along the coast of Albert county they are affected by faults which have in places brought up the Lower Carboniferous upon the middle portion and tilted the measures at a high angle.

The westward extension of the magnificent Joggins section touches the coast of New Brunswick only in its lower part ; the Productive measures, in which the Joggins coal seams are found, trending out into the bay, and being exposed on no part of the shore of the province. The general horizontality of the measures shows that the tremendous disturbances, which so profoundly affected the coal fields of Spring Hill and Pictou, did not extend in this direction, while the apparent absence of the coal measures over any portion of New Brunswick, in so far at least as can be at present determined, leads to the assumption that this area was already permanently raised above the sea.

Several outlying patches of the Middle Carboniferous are seen at points outside the central area. In character of sediments, they do not differ from those already described, and contain also a thin seam of coal. Among these may be mentioned that of Prince William, about twenty miles west of Fredericton and at Dunsinane, on the railway between Sussex and Petitcodiac.

Volcanic rocks. Extensive traces of volcanic activity are manifest at many points, especially in the Lower Carboniferous formation. Thick beds of ash rocks, conglomerates and breccias, together with various traps are found, not only around its border, but in isolated hills which, by the agency of denudation, now appear prominently through the grey sandstones of the Millstone-grit in the very centre of the coal district. In places, these trappean masses are apparently overflows from great dykes which have cut the lower beds of the Middle Carboniferous, and have altered these along the contact in the same manner as the sandstones of the Bay of Fundy have been affected by the trap outflows of the Triassic time. As, however, other areas of volcanic rocks undoubtedly belong to the basal beds of the system, it would appear that at least two periods of volcanic action occurred, unless we refer the later to the time of the outburst which affected the Bay of Fundy area.

Northern and western New Brunswick. Leaving for the present the consideration of the igneous rocks, which constitute an important element in the geology of this part of the province, we will now examine the structure of its northern and western portions.

The earlier investigations of Dr. Gesner and others have already been reviewed down to the year 1874. Upon the completion of the map of southern New Brunswick, the examination of these sections was at once commenced, the northern part by the author of this paper; the western by Prof. Bailey and Mr. W. Broad ; while the superficial geology of the district was entrusted to Mr. R. Chalmers.

R. W. Ells. L. W. Bailey. Work done in that area. Rep. Geol. Surv. 1879 to 1885. Reports on the former area appeared in 1879-80, 1880-81, by the writer, and on the latter by Prof. Bailey, 1882-83-84, and 1885. Investigations were also carried on for a short time in 1879, by Mr. G. F. Matthew, in Carleton county, the results of which were not published separately, but subsequently embodied in the report by Prof. Bailey on the area in question.

The difficulties encountered in tracing out the various formations, were far greater in this section than in that already described, owing to the fact that, with the exception of narrow belts of settled lands in the vicinity of the St. John River on the west, and along the shores of the Bay of Chaleurs on the east, and for a short distance inland along the course of the principal streams, the country is an entire wilderness ; the only means of access to its interior being by canoes along the various rivers which rise in close proximity to each other in the series of mountains which constitute the watershed. Of these streams the principal to the east, are the Restigouche, Nipisiguit and the branches of the Miramichi ; on the west, the Southwest Miramichi, the Tobique, and several other large branches of the St. John. From the explorations that have been thus carried on, the following systems have been clearly recognized :—

Pre-Cambrian ....Comprising gneisses, syenites, schists of various kinds and felsites.

Cambro Silurian ......Which, as in southern New Brunswick, may also comprise limited areas of pre-Cambrian, and even Cambrian rocks, at present • not separable.

Siluriah.................Limestones, slates, and sandstones, often highly fossiliferous.

Devonian.............. Principally developed about the Bay of Chaleurs, and in Gaspé peninsula, but also as patches overlying the Silurian rocks inland.

Carboniferous..........Lower or Bonaventure formation. Middle or Millstone-grit. Upper along the shore of the Gulf of St. Lawrence Gulf Shores.

Volcanic or Igneous Rocks.

In the study of the pre-Cambrian areas of northern New Bruns- Pre-Cambrian wick, no attempt has yet been made to separate the Huronian from of northern the Laurentian ; since, in the present largely inaccessible condition of R. W. Ells, the country, such separation must, of necessity, be very imperfect. Rep. Geol. Sur., Various rocks occur which approach very closely in character those of the typical Laurentian of St. John county, and which, apparently, form the lowest members of the pre-Cambrian series, being clearly distinguished from the upper or schistose portion. Of the former are the gneisses and syenites, with certain felsites, which are extensively developed about the head waters of the Nipisiguit River, whence they extend southwesterly, crossing the upper part of the south branches of the Tobique, but it is worthy of remark, that in all the areas of these rocks yet examined in this portion of the province, no traces of the crystalline limestones which make up so large a part of the upper division of the Laurentian of southern New Brunswick, has yet been found. On the Tête-à-gauche River, beds of graphitic, and to some extent, crystalline limestone occur, which resemble somewhat those of St. John, but their intimate association with blackish slates which belong to the Cambro-Silurian graptolitic series near the mouth of that stream, renders it probable that these belong to that horizon, and that their alteration is due largely to local intrusions of dioritic masses, in the same way as is seen in the alteration of the fossiliferous Silurian limestones into marble near the Bay of Chaleurs.

Generally, the pre-Cambrian rocks of this section present much resemblance to those of the recognized divisions already described. There is, however, probably, a greater preponderance of true felsites, and a smaller development of the ashy or volcanic portions of the Coldbrook and Coastal groups, but felsitic, chloritic and talcose schists, epidotic rocks, gneisses, &c., are common at many places throughout the area in question. They occupy an extensive tract of country extending diagonally across the northern portion of the province, from a short distance above the Forks of the Main South-West Miramichi, on the west, nearly to the mouth of the Jacquet River, on the Bay of Chaleurs. The greatest breadth of this area, on a line drawn across to the head of the Tobique River, is about forty-five miles, in which is included a breadth of twelve miles of intrusive granites. The principal pre-Cambrian area is overlapped by the Silurian on the east, a short distance after crossing the Tête-à-gauche River, on which stream its southern boundary is seen about sixteen miles from its mouth. The Silurian rocks here occupy an elongated basin, extending inland for some distance beyond the upper waters of the Upsalquitch River, bounded on the north by the prolongation of the felsite area, which can be traced continuously from the Tobique Lake across the latter stream, and along the north side of the Jacquet River, nearly to its mouth. On the upper part of the Upsalquitch, gabbros are associated with the gneisses that constitute the high hills in this vicinity ; but along the watershed which divides the streams flowing into the Gulf of St. Lawrence, viz., the various branches of the Miramichi and those into the Tobique, the pre-Cambrian rocks have been broken through by a great mass of granitic and syenitic rocks, precisely similar in character to those of Charlotte county. The southern margin is also for some distance bounded by similar granites, which are extensively exposed on the upper part of the Main South-West Miramichi and its tributaries.

On all the streams which flow east between the Nipisiguit and the Main South-West Miramichi, the pre-Cambrian rocks are easily recognized. Their intense degree of metamorphism, regional rather than local, distinguishes them from those of the overlying formations, while in many places there is a marked unconformity between them. The western boundary of the series is seen on the Nictor Lake, which is at the head of the Tobique River, in several islands and in the great peak, known as the Bald Mountain, whence a high chain of hills extends to the Right Hand Branch of the Tobique, crossing it a short distance below the Forks of the Campbell and Serpentine Rivers, and continues with a more south-westerly trend to the Forks of the Miramichi. On the Nipisiguit, its eastern boundary is seen a short distance below Indian Falls, not far from the 47-mile post on this stream. The north-eastern area is divided into three parts by basin-like overlaps of Cambro-Silurian and Silurian strata, presently to be described.

To the south, the pre-Cambrian is overlaid principally by rocks of the Cambro-Silurian system, and on the west by sediments of various age, up to the Lower Carboniferous.

The rocks of the Cambro-Silurian resemble those of the areas Cambro-already described under this head in previous pages. They consist Silurian. of slates, grey, red, and black, with quartzose sandstones, sometimes schistose, and where in contact with the granites, containing abundance of crystals, of staurolite, mica, &c. Certain bands of the red and green slates are persistent for long distances, and can readily be traced from their northerly terminus on the Bay of Chaleurs, southwesterly, into the county of York. In character, these slates of various colors, with their associated sandstones, resemble very closely the beds of the Levis and Sillery divisions of the Quebec group, and this resemblance is strengthened by the occurrence of graptolites, in some of the graphitic layers, similar to those found in that group. As in the southern portion of the province, certain areas of highly metamorphic rocks also occur which lithologically closely resemble pre-Cambrian, but as in the case of the more southern area, these are so intimately associated with other sediments, as to render their separation impossible for the present.

The rocks now considered of Cambro-Silurian age are in great part Formerly those described by Drs. Gesner and Robb as Cambrian, while in the described as Cambrian by

4

map accompanying the Acadian Geology, they were comprised under the term Lower Silurian. To the south-west, they occupy a large
portion of the county of York, whence they extend into the adjoining state of Maine. In this direction, they are directly overlaid by
Carboniferous sediments of the central basin, the Devonian and
Silurian being entirely concealed, or, at best, represented only by
detached outliers of very limited extent. Their north-west outline
crosses the province boundary into Maine, about two miles south of
what is known as the Monument, at the source of the St. Croix
River, whence, trending north-easterly, it crosses the St. John River,
a little to the north of Woodstock, and extends nearly to the head
waters of the Tobique.

Areas of limestone which are often highly crystalline and micaceous are found as an integral portion of the system in the western
area, near Canterbury. These resemble in some respects the crystalline limestones of the Laurentian, and are associated with quartzites and schistose rocks. The occurrence also of somewhat similar
limestones in the eastern area as a part of the same formation has
already been referred to in the remarks on the pre-Cambrian.

While then the Cambro-Silurian rocks have a somewhat extensive
development in this section of the province, their bulk is much
reduced by the presence of large masses of syenite and granite of
undoubtedly much later date. These have penetrated the surrounding beds into which large dykes and veins are intruded in all directions, while often large pieces have the appearance of being torn from
their original position, and now held in the igneous mass. Their
action on the slates and sandstones is marked, not only by the gene-
ration of crystals of various kinds, but by a general alteration of the
contiguous beds into schists and gneisses containing mica.

Two principal areas of these rocks are seen, the more southern
flanking the south side of the pre-Cambrian axis of the interior, and
continuing to the shore of the Bay of Chaleurs, north of Bathurst,
the other on the north-west side of the axis, terminating, as described,
near the head waters of the Tobique. These areas are distinctly unconformable to the underlying series. They are not, as a rule, rich in fossils, but at several points in the southern belt, notably on the Tête-à-
gauche and Miramichi rivers, different forms are found, including
brachiopods as well as graptolites. These, while in many cases too
indistinct for perfect determination, present features more nearly allied
to Cambro-Silurian forms than to those of any other horizon, but in

the western belt, more especially on the North Branch of the Becca- Prof. L. W.
quimec River, in Carleton, a wonderfully mixed fauna is found in western New
a somewhat limited space. At one point, near Shaw's mill, on Brunswick,
Rep. Geol. Sur.
this stream, strata-holding fossils which appear to belong to this 1885.
system are apparently interstratified with others containing Silurian
forms, while in close proximity are beds filled with remains of *Psilo-*
phyton. There would, therefore, in this locality, appear to be three
systems represented, of which the Silurian forms seem to be near the
base or junction with the Cambro Silurian. This peculiar admixture of
so many different horizons can probably best be explained on the hypo-
thesis of intimate infolding and subsequent denudation by which
narrow crests of older ridges are exposed. In addition to the fossils
recently found at this place, several small outlying patches were noted,
and first referred to by Mr. C. Robb (Rep. Geol. Sur. 1869), and
others subsequently discovered by Mr. McInnes, in northern York ;
but these appear to be more closely allied to Lower Helderberg forms,
and consequently quite distinct from the rocks of the principal Cambro-
Silurian area upon which they rest as limited outlying patches.

The view taken of this portion of the rocks in northern New Similarity of
these rocks to
Brunswick, by Professor Hind, as representing the Quebec group of portions of the
Canada (see his Rep. 1865), has thus been fairly sustained by later as stated by
investigation, both on grounds of lithology and palæontology. The Prof. H. Y.
Hind. Report
similarity also of the fossils of the Beccaquimec area to those of the N.B. Gov., 1865.
Trenton group, to which horizon, a portion, at least, of the fossiliferous Quebec group belongs, is also evident, and leads to the conclu-
sion, that a repetition of these rocks, both in their fossiliferous and
metamorphic stages, is found in this section of the province.

But by far the most extensive of the older geological systems in Silurian.
this area is the Silurian. This occupies the entire country along
the St. John River, above Woodstock, extending far into the adjoining province of Quebec, together with the greater part of the valley of
the Tobique, where it is, however. to some extent overlapped by
Lower Carboniferous sediments. Thence it extends to the Bay of
Chaleurs ; occupying, with the exception of the pre-Cambrian belt of
Jacquet River, and sundry areas of diorite and trappean rocks the
remaining portion of the province to the north, and including the
valley of the Restigouche and its tributaries, as well as a large portion of the Gaspé Peninsula, where it rests upon the southern flank
of the Quebec group.

The strata of this system are thrown into a series of anticlinals, the axes of which are well exposed on the various streams, and the beds are in places highly fossiliferous, the different formations from the Niagara to the Lower Helderberg, both inclusive, being already recognized. They are frequently penetrated by dykes and masses of trap, often of large extent, some of which, as at Dalhousie, are intercalated sheets between the fossiliferous limestone and shales, which have been altered along the contact. The Silurian here contains traces of plant stems, which are probably the oldest found in the province, and are associated with distinctly Silurian forms. The same association of plants in Silurian strata is found in the Gaspé limestone series near Gaspé Basin, and alluded to by Sir Wm. Dawson in " The Fossil Plants of the Silurian and Dev., 1871." It is also possible that the plant stems noted on the Beccaguimec may be in the upper beds of the Silurian, though their aspect is at this place, more markedly Devonian.

Devonian.

L. W. Bailey on Carleton and Victoria counties. Rep. Geol. Sur. 1885.

The rocks of the Devonian system occupy but limited areas in this section, and are, for the most part, confined to the vicinity of the upper portion of the Bay of Chaleurs. Several small outliers, have, however been recognized in Carleton and Victoria counties, the fossils of which would place them near the base of the series. Of these, the former is found near the junction of the Beccaguimec, with the St. John, where certain black shales are exposed in a narrow band containing abundant remains of *Psilophyton princeps*, a characteristic Devonian form (see Rep. Geol. Sur., C. Robb, 1870-71). This area is largely concealed by Lower Carboniferous conglomerate and sandstone, which occupy a basin formed by the branches of the first-named stream.

These rocks were in 1874 (see Rep. Geol. Sur. 1874-75, R. W. Ells,) on the evidence of the contained plants, regarded as belonging to the same horizon, but later detailed examinations about the upper part of the stream, by Messrs. Bailey, Matthew and McInnis, tend to establish a later age for the conglomerate and associated beds. (See Rep. Geol. Sur. 1883-84, L. W. Bailey.) Devonian rocks, however, underlie these, since along the northern boundary, on the North branch, certain fossils are recognized which appear to represent the upper part of the Gaspé limestone series.

R. W. Ells, northern New Brunswick. Rep. Geol. Sur. 1879-80.

In the section along the Upsalquitch River, a branch of the Restigouche from the south, a basin of Devonian sandstone and shales with characteristic plants was observed to rest unconformably upon Silurian rocks. Its extent inland could not be traced, as the surrounding country is generally low and densely wooded. As at other points the sediments were intersected by trappean masses, whose metamorphic action was quite evident and proved their more recent age.

On the Lower Restigouche areas of Devonian rocks are found on both sides of the stream which forms the dividing line between the provinces of New Brunswick and Quebec. These, on the south side, extent from a point three miles above Campbellton, at intervals to within a couple of miles of Dalhousie. The shales and sandstones contain plant stems, descriptions of which are given in the report of R. W. Ells, northern New 1879. In the vicinity of Campbellton, also, beds of brecciated Brunswick. Rep. Geol. Sur limestone or calcareous breccia have yielded a comparatively rich 1879-80. fauna of Devonian fishes such as *Cephalaspis Coccosteus*, etc., representing the lower part of the system, while on the side of the river, opposite Dalhousie, other beds contain an abundance of fossil fishes, which have been described by Mr. Whiteaves, and appear to indicate its middle or upper portion. It is of interest to note that as Fossil fishes near Dalhousie early as 1842 these remains were recognized by Dr. Gesner, who, first noted by however, regarded them as reptilian in their character. He supposed Gesner in 1842. the containing beds, from their lithological aspect were portions of the New Red Sandstone or Carboniferous formations, and it was not till 1879, that these interesting fossils were rediscovered by the writer and the true position of the beds established. The Devonian of this locality occupies a shallow synclinal. The strata are penetrated by trap dykes, two periods of eruption being evident from the fact that the lower beds at Campbellton, in which the fishes were found are composed largely of trappean debris, as also from the presence of pebbles of trap in the conglomerates elsewhere, and the occurrence of fossiliferous strata overlying trappean ridges. Later intrusions of volcanic matter through the newer members of the system, both here and at various points in the Gaspé Peninsula, are also common.

The Carboniferous system is represented in northern and eastern Carboniferous of northern New Brunswick, principally by the upper portion of the Lower forma- and eastern tion, styled the Bonaventure and the lower or Millstone-grit portions Brunswick. of the middle division. The development of the latter over the great inland basin has already been referred to and but little more need be said concerning it. Along the south side of the Bay of Chaleurs a thin seam of coal occurs at several points, which has the same general character as in the interior. On the island of

Shippigan, and on the mainland north of Tracadie, soft, red micaceous sandstone is seen which probably represents a part of the
Upper Carboniferous, but these areas are confined to a narrow fringe
along the shore.   A similar thin seam of coal is found on several of
the branches of the Miramichi, and gives strong evidence that the
formation has no great thickness at any point.   The area is traversed
by several low anticlinals, of which four principal ones are recognized and described in Rep. Geol. Sur. 1882–83.

R. W. Ells,
Rep. Geol. Sur.
1882-83.   The more northerly of these extends between Bathurst and the
Miramichi River, where it forms a ridge running north-easterly with
an elevation of between 500 and 600 feet.   The second extends from
the head of Grand Lake to the vicinity of Richibucto Head on
Northumberland Strait.   This brings up the Devonian rocks of
Coal Creek.

The third passes to the north of Moncton, indicated by the
supposed pre-Cambrian ridges of Indian Mountain, and reaches the
shore a few miles north of Shediac ; while the fourth, which affects
the south-eastern area only, is well seen in the Aulac ridge which
extends to Bay Verte, and thence in a low rise runs through the Tormentine Peninsula to its extremity.   The basin of Middle Carboniferous rocks is underlaid along nearly its entire boundary on the north,
west and south by the Lower Carboniferous.   At one or two points
however, notably on the Dungarvon and Renous Rivers and on
either side of the St. John River, west of Fredericton, the lower
members are concealed by the overlap of the Millstone-grit, directly
upon the Cambro-Silurian.   The volcanic portion of the Lower Carboniferous though considerably developed in the counties of York
and Victoria, is apparently absent from the northern area where the
rocks are sandstones and shales with conglomerates, which in the
southern part of the province, constitute the upper members of the
gypsiferous division.   They are well displayed along the south coast
of Gaspé, at intervals to the extremity of that peninsula, where they
received the name of the Bonaventure formation, as well as at several
points on the shore north of Bathurst.   In Carleton county also,
considerable outliers of Carboniferous rocks, representing the lower
and possibly some portion of the middle division, are found north of
Woodstock and on the branches of the Beccaguimec River.

These overlie Devonian and Silurian sediments already described,
which in turn rest upon rocks of both Cambro-Silurian and pre-Cambrian age, and conceal the contacts of the several systems for a con-

siderable distance. A considerable area principally of the gypsiferous portion is found on the Tobique, and recognized as long ago,
as the time of Dr. Gesner. Certain not clearly defined grits and
sandstones, grey in their upper part, may indicate the existence of
a patch of Millstone-grit in this locality.

On Heron Island which is in the Bay of Chaleurs, about nine
miles southeast of Dalhousie, the rocks of the Bonaventure formation
are well developed. The shales here contain remains of plants which
have never yet been described, while the sandstones shew impres-
sions of reptilian footmarks which are the only ones yet found in
New Brunswick of this age, though somewhat similar tracks have
been recognized at several points in Nova Scotia in strata of not
much higher horizon.

IGNEOUS ROCKS.

The principal igneous rocks, which have not already been sufficiently described in connection with the various formations, are
the intrusive granites and syenites and those of the newer trappean
areas.

Of these the former are much the more important, not only from
their very considerable extent but from their economic value. They
have been roughly outlined from the time of the earliest report on
the geology of the province, but it is only within the last few years,
that the details, more especially of the northern area, which is much
the larger, have been studied.

Generally speaking, these two great areas enter the province near
its south-west corner from the adjoining state of Maine. Along the
border in Charlotte and York counties, they are separated by a
considerable extent of slates already described. The southern belt
extends entirely across the former county, and occupies a large part
of western Kings and Queens, reaching, with some interruptions, to
the St. John River. Further east, in the latter county and in Westmorland, isolated outcrops protrude through Carboniferous sediments revealed by the denudation of the latter. That these are of
earlier age than the overlying beds is proved by the debris of the
granites being found in the lowest member of the Lower Carboniferous formation. Similar areas of granite are also associated with
the pre-Cambrian of the southern part of the province, but these are
comparatively limited as compared with the principal granitic mass.

In character these rocks are very similar, being generally reddish and moderately coarse grained, often with large crystals of felspar. In places, however, the texture is fine and the color grey.

The northern area, while of greater extent, presents a similar aspect. Crossing from Maine through the chain of the St. Croix lakes, which form the boundary for some distance, it enters the province in a belt more than twenty miles in width, and extending north-easterly, crosses the St. John River, midway between Fredericton and Woodstock. Fine sections are afforded by the river, which cuts directly through the belt, shewing well the intrusive character of the rock by the number and nature of the dykes, which are sent off in all directions into the adjoining slates, as well as by the distinct local metamorphism, due directly to the presence of the granitic mass. Crossing the River St. John, the granite subdivides into two portions, the more southerly of which terminates near the New Brunswick railway, while the northern band crosses that line and continues to a total distance of thirty miles from the river. North of this, to the Bay of Chaleurs, the granites occur in three distinct areas, of which the two central are of large extent, and are, for the most part, associated with pre-Cambrian rocks, though also penetrating strata of Cambro-Silurian age, on the Miramichi River. On this stream, the alteration along the contact, both of the slates and granite, is well seen, the latter, for several feet, becoming fine-grained and whitish in color, while the former are in places shattered and contain crystals of various kinds. The second area occupies a great breadth of country about the head waters of the South Branch of the Nipisiguit, and on the North-west Miramichi ; forming an exceedingly hilly and broken surface, containing the highest peaks of the province. The third area occurs on the lower part of the Nipisiguit near Bathurst, extending for some twelve miles up from its mouth, but concealed on the lower portion for three miles by Lower Carboniferons beds, the base of which is made up of granitic debris.

From a consideration of the various points of contact, the intrusive character of these granites is very clearly established ; since, in no other way, can the peculiar phenomena seen be accounted for. There is, however, a marked difference in the character of the metamorphism resulting from the granite, as compared with that produced by the intrusion of the dioritic or trappean masses. In the former, the alteration is more gradual, and extends over a much wider area, as though continuing for a considerable period, and probably under great pressure. In the latter case, the rocks, in contact have frequently a baked or porcelainized aspect, as though exposed to a quicker but not so prolonged a heat. This may, perhaps, be more clearly expressed by regarding the granites as *intrusive* rocks proper, which have not reached the surface at the time of their intrusion, but cooled beneath the surface, and subsequently exposed by denudation, while the latter may be held as rather *extrusive*, coming to the surface along direct fractures or lines of bedding, and cooling rapidly.

Of the other kinds of intrusive rocks, many of the diorites and felsites doubtless are, as heretofore described, integral parts of the older formations ; for while the metamorphosing action of the granite is seen in beds of Silurian age, it is evident that its intrusion must have been subsequent to that period ; while, as no pebbles or debris are found in any formation older than the Devonian, the age of these great masses must not be far from the beginning of that era. Many of the older volcanic rocks of the pre-Cambrian areas have an earlier date than the overlying Cambrian strata, since their debris enters largely into the composition of the basal beds of that system.

The areas of serpentine in the province are too limited to form any particularly distinctive feature. Sufficient evidence, however, exists to indicate its connection with the igneous rocks, probably as a product of alteration, as seen in the serpentinous diorites of western Charlotte county and in the pure serpentine dyke near Pisarinco.

The trappean rocks, which are largely developed along the lower Restigouche, and along the upper part of the Bay of Chaleurs, are, like the granites, for the most part of Devonian age. In places, dykes of considerable size, cut directly through sandstones and shales of this age, or throw them upwards into low anticlinals, in the same way as the traps of the Bay of Fundy have affected the Triassic sandstone of that locality. In places, also, the conglomerates of the period, are largely made up of trappean debris. It would thus appear that at least two periods volcanic eruption occurred in this region, of which the earlier was probably the more extensive ; since by it the huge mountains of Dalhousie, Campbellton, and the range along the north shore of the Restigouche to Tracadigash, were brought into their present position, upon the flanks of which nearly horizontal beds of Devonian age have been deposited at various points.

There yet remains to be considered the great masses of felsitic rock of various ages, both of the southern portion of the province and of the great areas of the northern division. That the great bulk of these are true volcanic products is evident from their nature. Their character as amygdaloids, agglomerates and ashbeds, together with the existence of highly crystalline felspar porphyries, trachytes, rhyolites and similar rocks, clearly establishes their eruptive origin. That they are, however, of different ages, is plain, since while some of these are undoubtedly of pre-Cambrian age, and form what is known as the volcanic portion of the Huronian system, others are intimately associated with Lower Carboniferous strata, either as great masses or as interbedded sheets. These latter are, however, as a rule more earthy than the older series, and their mode of occurrence clearly indicates their later age.

A large area of felsite occurs on the Tobique River, the age of which is somewhat doubtful. Much of it is highly crystalline and porphyritic, and resembles the old pre-Cambrian felsite of the interior; other portions are ashy. It is separated from the main area by a band of Cambro-Silurian slates, which may only occupy a basin shaped valley in the older rocks, but the impenetrable character of much of the country renders it at present almost impracticable to determine its true horizon, and it is possible that it may pertain to the Lower Carboniferous outlier of this locality, with the rocks of which it appears to be intimately connected. Other areas of epidotic and dioritic rocks of uncertain position occur in the vicinity of the St. John River, near Woodstock, which are associated with Cambro-Silurian strata.

SUPERFICIAL GEOLOGY.

The superficial geology of the province has, since the report of Prof. Hind (1865), been more particularly studied by Messrs. Matthew and Chalmers, whose reports embody not only their own observations on the subject, but many of the notes collected by the other explorers in this field.

Prof. H. Y. Hind, Report Geology of New Brunswick, 1865.

That by Prof. Hind furnishes much valuable information concerning the origin of terraces, raised beaches and lake basins; with a list of glacial striae and ice grooves from which he infers that the ice might have had a thickness of 2,000 feet, as indicated by the markings found on the tops of some of the mountains. The varying courses of the striae are accounted for on the grounds that the direction of the ice sheet was influenced, to some extent, by the direction of the valleys and other leading topographical features. This pecu-

liarity of local glaciers has been discussed at greater length in recent
reports by Mr. Chalmers. The various lake basins were held to be
due to the ploughing out of the softer portions of the underlying rocks
through the agency of these glaciers.

The reports of Mr. Matthew go much more extensively into detail. Mr. G. F.
In these the superficial deposits are arranged under three heads, viz. : Matthew,
superficial
1st. Boulder clay or till, unmodified glacial drift, constituting the geology of
southern New
lowest member. Brunswick.
Rep. Geol. Sur.·
2nd. Stratified sand and gravel, Syrtensian deposits, formed by 1877-78.
marine action, and representing the remains of old shoals and banks.

Leda clay, estuarine deposits.

Saxicava sand and raised beaches, littoral deposits.

3rd. Modern alluvium, shell marls, peat, etc.

The peculiarity of the boulder clay is its unstratified arrangement
and its intermingling of sand, clay and stones which are often striated.
As a general rule, it was observed that the boulders were, for the most
part, local, and only at rare intervals were stones found that had come
from any considerable distance. Where observed, in the southern part
of the province, their course was generally from north to south or southeast, more especially in the south-western and central portions, and
good instances of upward transport were seen by their presence on
the sides or tops of mountains, 500 feet or more above their starting
point.

The color of the boulder clay was found to be affected in great
measure by the color of the rocks whence it was derived, as might
naturally be supposed, more especially where these rocks are soft
sandstones or calcareous shales, since the hard rocks resisted more
effectually the degrading effects of the ice sheet. Wherever this clay
is removed, the strata beneath are, as a rule, found to be rounded
and scored, shewing that the deposit of the clay was subsequent to
the smoothing of the rock. Two principal directions are visible in
the striæ. Thus to the west of St. John the prevailing course is to
the south of east, while to the east of that city, it is south-westerly,
following to a great extent the principal hill features.

The beds of the second or Syrtensian group differ from the preceding in their stratified character, though there are, at certain points
on the coast, indications of a gradual passage from the upper members of the boulder formation, into the lower part of the stratified
group. The various beds of the latter are arranged by Mr. Matthew
in four principal forms, viz. :—

The weather shoal, forming under the north side of the hill, or at the entrance to a valley.

The lee shoal, forming behind a hill, or at the outlet of a valley.

Centre shoals, formed in open spaces or enlargements of valleys, or on the higher lands.

Horsebacks and escars, formed of gravel and sand along valleys or ridges as connecting neighbouring hills, or opposing slopes of valleys.

The Leda clay may be regarded as an upper portion of the preceding group, since, in the southern part of the province, the beds of the former graduate upward into the latter. It may, however, be properly restricted to deposits of clay, which generally rest upon the gravels of the Syrtensian group, and are generally distinguished by the presence of organic remains, in which, in addition to shells of various kinds, the remains of a *Phoca* and a *Beluga* have been found nea the coast of the Bay of Fundy. This deposit is well recognized, not only in the southern part of the province, but in the Bay of Chaleurs area, both in New Brunswick and along the Gaspé shore, from which a very considerable collection of fossil remains have been obtained. The deposit, however, changes its character in different places, from the admixture of sandy beds, and at times it is very difficult to distinguish between the Syrtensian, the Leda and the Saxicava, owing to their occasional similarity.

To the latter or Saxicava sand are attributed the upper members of the modified drift only. It is generally devoid of organic remains, having in southern New Brunswick, produced only two species, a *Mya* and a *Macoma*. It was deposited in gradually shoaling water, as the land slowly rose from the sea, and, in this way, terraces of different heights were formed along the coast and river valleys. These terraces, more especially as seen along the St. John and other rivers in northern New Brunswick, have been recently described in detail by Mr. Chalmers, whose papers on the surface geology of these portions of the province, are among the most important yet published on the subject. In many of these terraces, the three sub-divisions of the modified drift are seen, viz. :—the Syrtensian gravel and sand at the bottom, the Leda clay in the central part, and the Saxicava sand at the top.

Mr. R. Chalmers. Superficial Geology. Rep. Geol. Sur. 1882-83-84.

On the St. John River, a number of sections were made, particularly of the part between Woodstock and the St. Francis River, as well as along its tributaries, in some of which no less than seven distinct terraces were observed, with a total elevation for the upper of not far from 200 feet above the present level of the stream. The highest terraces are composed of stratified gravel and sand, with water-worn pebbles; the lower contains the same materials as the upper, but are generally more water-worn, and have local beds of clay and silt. The materials of the kames or escars and terraces are the same, and undoubtedly, derived from the same source. Since it has been conclusively established that the valley of the St. John River is of pre-glacial age, the mode of occurrence of these deposits is an exceedingly interesting question, and as Mr. Chalmers has evidently given the subject much study, and his views seem to satisfy most of the requirements of the case, more fully than any previously advanced, with the proviso that they apply equally well to other river valleys, they are here presented.

" The river valleys, at the close of the glacial period, became very generally choked with drift, which, forming dams, would hold back the water, and constitute a series of lake basins along the river course. These dams, in some places, were from 150 to 200 feet above the present level, and the rivers would, therefore, begin to flow at that height above their present beds. The gradual re-excavation of the drift would, therefore, as it went on, by the transportation to lower levels, result in the formation of the terraces which would thus mark different levels of the river by the deposition of the materials from higher levels along the border of the lake-like expansions, and along the sides and bottoms of the current, which flowed in and through them ; thus, by successive accumulation, forcing the channel from side to side as erosion and deposition went on.

" The kames are composed of similar sand and gravel as the terraces, and may be the remains of these left by denudation of beds once surrounding them, and of which they formed a part. Their bases are often composed of till or boulder clay."

The action of the glaciers, in following already defined depressions, is well seen at many points over the entire area of the province. Occasionally, two or more sets of striæ occur, which often have courses at marked angles to each other.

Of these, the newer will almost invariably be noticed to run in the direction of a valley or, when near the coast, the course of some fiord or estuary. Excellent examples of this are seen in the vicinity of Sackville and Amherst, where the ice grooves and striæ follow directly the depression between Bay Verte and the head of the Bay of Fundy, as also at Memramcook and Dorchester, where they are observed to follow the principal flexures of the Memramcook River. This peculiarity of direction is also noticed in both the adjoining provinces of Quebec and Nova Scotia. The older set of markings are however observed to keep a quite uniform direction, regardless of opposing hills, as if in many cases, the propelling power of the glacier drove it resistlessly forward. If then we admit the existence of a universal ice sheet, which, in some parts of eastern Canada· appears to be quite clearly established, we must also admit a second and possibly a third ice era, during which local glaciers were shed from the height of land in whichever direction the most favorable outlet was presented, the course of which was largely affected by local conditions of configuration. In the northern area, such evidences of local glaciers are visible in the striæ which follow the outlines of the lower Restigouche and its tributaries. In Gaspé, also, local glaciers were undoubtedly shed in either direction from the tops of the Shickshock mountains, modified by the lower hill ranges which lie nearer the coast. In Nova Scotia, striæ on the north side of the Cobequids point towards the shore of Northumberland Strait, while on the south side, they have a westward course in the line of the Minas Basin and Channel.

The divergence of river beds is conclusively proved at several points, though whether these changes were all due to causes subsequent to the glacial period, may be questioned. Thus, at the Grand Falls of the St. John River, 225 miles from the mouth, the present channel below the pitch forms a wonderful gorge, cut through the Silurian rocks for nearly a mile to a depth of over one hundred feet ; the old channel being blocked, at what is now the town of Grand Falls, by a great accumulation of clay and other drift material. So also, near its junction with the St. John, the old valley of the Tobique has been dammed and a new channel excavated through Silurian beds for nearly the same distance. The time necessary to produce a gorge of such a length in the comparatively hard rocks of these localities must have been considerable. Whether such was the cause of the present outlet of the St. John River has not been conclusively proved, but appears highly probable, since huge dams of drift block up what appears to be an old outlet of the river in the direction of Manawagonish Beach, a short distance west of its present mouth.

The views of Prof. Hind (see Report of 1865), as to the glacial origin of many of the lake basins, do not, in all cases, seem to be fully sustained by the later investigations of Mr. Chalmers. The latter gentleman finds the depressions of some of the larger lakes in the southwestern part of the province to be pre-glacial in the same way as the valley of the St. John River. The subsequent action of glaciers and glacial drift has, by heaping up moraines, modified existing conditions very greatly, either by forming entirely new basins, or changing very considerably those existing prior to the ice age. This peculiarity of morainic lake basins is well illustrated in many of the lakes of the St. Croix chain, along the boundary between New Brunswick and Maine.

Changes of level of the land are well shown by the presence of beds of marine or Leda clay at considerable elevations, reaching, in some places, several hundred feet above the present sea level. It is also proved by the presence of old sea beaches at intervals along the coast of the Bay of Fundy, and now removed several miles from the existing shore line. That such changes of level have taken place in comparatively recent times is evidenced by the presence of tree stumps in the marshes about the upper part of this bay, now some twenty òr thirty feet below high tide mark, which are found not only on the flats between Amherst and Sackville, along the shore, but in the several canals which have been cut for the purpose of improving the inner marsh. The partially submerged shell heaps on the coast of Charlotte county, which mark the sites of old Indian encampments, and presumably at one time removed beyond the action of the sea, also proves the gradual encroachment of the water, as well as the fact that these shell heaps are at a considerably lower level than when they were formed. On the other hand, it is supposed by Prof. Hind and others that the shores of the Gulf of St. Lawrence are now gradually rising, as is evidenced by the shoaling of the waters in the harbors and estuaries of the eastern coast, at the entrance to the great Miramichi Harbor and the basin at Bathurst, since a marked difference in the depth of the water is now noted when compared with the observations recorded within the last eighty or one hundred years. Some of these apparent changes of level may, to a certain extent, be due to the filling up of channels by sediment brought down by the rivers and redistributed by the action of the tides.

It might be supposed, by some persons, that the varied opinions expressed at different times concerning the true interpretation of the geological structure of the province would have a tendency to reflect somewhat unfavorably upon the character of the work done by the several observers in this field. A moment's consideration of the subject will, however, serve to correct this impression; since it must be remembered that, in the elucidation of the structure of any country, the discovery of new facts, from time to time, must ever lead to changes in the interpretation of the various problems presented.

The earliest workers had the misfortune also to labor at a time when the science of geology was, comparatively speaking, in its infancy. With the increase of our scientific knowledge, enlarged views will be unfolded, while the peculiar bias pertaining to each individual must, of itself, ever give abundant cause for difference of opinion. In the study of any subject the object primarily to be attained should be the truth, sinking personal feeling for the general good.